THE AL QAEDA CONNECTION

Paul L. Williams

CONNECTION

International Terrorism,
Organized Crime,
and the
Coming Apocalypse

 Prometheus Books

59 John Glenn Drive
Amherst, New York 14228-2197

Published 2005 by Prometheus Books

Inquiries should be addressed to
Prometheus Books
59 John Glenn Drive
Amherst, New York 14228–2197
VOICE: 716–691–0133, ext. 207
FAX: 716–564–2711
WWW.PROMETHEUSBOOKS.COM

09 08 07 06 05   5 4 3 2 1

Library of Congress Cataloging-in-Publication Data

Williams, Paul L., 1944–
    The Al Qaeda connection : international terrorism, organized crime, and the coming apocalypse / by Paul L. Williams.
        p. cm.
    Includes bibliographical references and index.
    ISBN 1–59102–349–1 (hardcover : alk. paper)
    1. Qaida (Organization) 2. Transnational crime. 3. Drug traffic. 4. Nuclear weapons. Mafia—Italy—Sicily. 6. Mafia—Russian (Federation)—Chechnia.
I. Title.

HV6433.M52Q35 2005
303.6'25—dc22

                                                                    2005013880

Printed in the United States on acid-free paper

For Patricia, the love of my life

CONTENTS

PREFACE

An Interview with Osama bin Laden

This was the first interview given by Osama to any journalist after the September 11 attacks in New York City and Washington, DC. Hamid Mir, the correspondent for the Pakistani newspaper Dawn, *was taken blindfolded in a jeep from Kabul on the night of November 7 to an "extremely cold" place, where he could hear "the sound of antiaircraft guns firing away." After a wait of several hours, bin Laden arrived with about a dozen bodyguards and Dr. Ayman al-Zawahiri, who also answered questions. The interview was carried by* Dawn, *BBC, and the Reuters news service. It was also aired by Al-Jazeera, the Arabian news channel.*

Hamid Mir: After the American bombing on Afghanistan on October 7, you told the Al-Jazeera TV network that the September 11 attacks had been carried out by some Muslims. How did you know they were Muslims?

Osama bin Laden: The Americans themselves released a list of the suspects of the September 11 attacks, saying that the persons named were involved in the attacks. They were all Muslims, of whom fifteen belonged to Saudi Arabia, two were from the UAE, and one from

Egypt. According to the information I have, they were all passengers. *Al-Fatihah* [the Exordium] was held for them in their homes. But America said they were hijackers.

HM: In your statement of October 7, you expressed satisfaction over the September 11 attacks, although a large number of innocent people perished in them, hundreds among them were Muslims. Can you justify the killing of innocent men in the light of Islamic teachings?

OBL: This is a major point in jurisprudence. In my view, if an enemy occupies a Muslim territory and uses common people as [a] human shield, then it is permitted to attack that enemy. For instance, if bandits barge into a home and hold a child hostage, then the child's father can attack the bandits and in that attack even the child may get hurt.

America and its allies are massacring us in Palestine, Chechnya, Kashmir, and Iraq. The Muslims have the right to attack America in reprisal. The Islamic *shariah* [law] says Muslims should not live in the land of the infidel for long. The September 11 attacks were not targeted at women and children. The real targets were America's icons of military and economic power.

The Holy Prophet (Peace be upon him) was against killing women and children. When he saw a dead woman during a war, he asked why was she killed? If a child is above thirteen and wields a weapon against Muslims, then it is permitted to kill him or her.

The American people should remember that they pay taxes to their government, they elect their president, their government manufactures arms and gives them to Israel, and Israel uses them to massacre Palestinians. The American Congress endorses all government measures and this proves that the entire America is responsible for the atrocities perpetrated against Muslims. The entire America, because they elect the Congress.

I ask the American people to force their government to give up anti-Muslim policies. The American people had risen against their government's war in Vietnam. They must do the same today. The American people should stop the massacre of Muslims by their government.

HM: Can it be said that you are against the American government, not the American people?

OBL: Yes! We are carrying on the mission of our Prophet, Muhammad (Peace be upon him). The mission is to spread the word of God, not to indulge massacring people. We ourselves are the target of killings, destruction, and atrocities. We are only defending ourselves. This is defensive *jihad* ("holy war"). We want to defend our people and our land. That is why I say that if we don't get security, the Americans, too, would not get security.

This is a simple formula that even an American child can understand. This is the formula of live and let live.

HM: The head of Egypt's Jamia al-Azhar has issued a *fatwa* (edict) against you, saying that the views and beliefs of Osama bin Laden have nothing to do with Islam. What do you have to say about that?

OBL: The *fatwa* of any official Aalim [a person with knowledge of Islamic teaching who is not a cleric] has no value for me. History is full of such *ulama* [scholars] who justify *riddah* [apostasy], who justify the occupation of Palestine by the Jews, who justify the presence of American troops around Harmain-e-Sharifain [the holy cities of Medina and Mecca]. These people support the infidels for their personal gain. The true *ulama* [possessors of religious knowledge] support the *jihad* against America. Tell me if Indian forces invaded Pakistan what would you do? The Israeli forces occupy our land and the American troops are on our territory. We have no other option but to launch *jihad*.

HM: Some Western media claim that you are trying to acquire chemical and nuclear weapons. How much truth is there in such reports?

OBL: I heard the speech of American President Bush yesterday [October 7, 2001]. He was scaring the European countries that Osama wanted to attack with weapons of mass destruction. I wish to declare that if America used chemical or nuclear weapons against us, then we may retort with chemical and nuclear weapons. We have the weapons as deterrent.

HM: Where did you get these weapons from?

OBL: Go to the next question.

HM: Demonstrations are being held in many European countries against American attacks on Afghanistan. Thousands of the protesters were non-Muslims. What is your opinion about those non-Muslim protesters?

OBL: There are many innocent and good-hearted people in the West. American media instigates them against Muslims. However, some good-hearted people are protesting against American attacks because human nature abhors injustice.

The Muslims were massacred under the UN patronage in Bosnia. I am aware that some officers of the State Department had resigned in protest. Many years ago the US ambassador in Egypt had resigned in protest against the policies of President Jimmy Carter. Nice and civilized are everywhere. The Jewish lobby has taken America and the West hostage.

HM: Some people say that war is no solution to any issue. Do you think that some political formula could be found to stop the present war?

OBL: You should put this question to those who have started this war. We are only defending ourselves.

HM: If America got out of Saudi Arabia and the Al-Aqsa mosque was liberated, would you then present yourself for trial in some Muslim country?

OBL: Only Afghanistan is an Islamic country. Pakistan follows the English law. I don't consider Saudi Arabia an Islamic country. If the

Americans have charges against me, we, too, have a charge sheet against them.

HM: The Pakistani government decided to cooperate with America after September 11, which you don't consider right. What do you think Pakistan should have done instead of cooperating with America?

OBL: The government of Pakistan should have the wishes of the people in view. It should not have surrendered to the unjustified demands of America. America does not have solid proof against us. It just has some surmises. It is unjust to start bombing on the basis of those surmises.

HM: Had America decided to attack Pakistan with the help of India and Israel, what would we have done?

OBL: What has America achieved by attacking Afghanistan? We will not leave the Pakistani people and the Pakistani territory at anybody's mercy.

We will defend Pakistan. But we have been disappointed by General Pervez Musharraf. He says that the majority is with him. I say the majority is against him.

Bush has used the word "crusade." This is a "crusade" declared by Bush. It is no wisdom to barter off the blood of our Afghan brethren to improve Pakistan's economy. He will be punished by the Pakistani people and Allah.

Right now a great war of Islamic history is being fought in Afghanistan. All the big powers are united against Muslims. It is *sawab* [a heavenly reward] to participate in this war.

HM: A French newspaper has claimed that you had kidney problem and had secretly gone to Dubai for treatment last year. Is that correct?

OBL: My kidneys are all right. I did not go to Dubai last year. One British newspaper has published an imaginary interview with an

Islamabad dateline with one of my sons who lives in Saudi Arabia. All this is false.

HM: Is it correct that a daughter of Mullah Omar is your wife or your daughter is Mullah Omar's wife?

OBL: (Laughs). All my wives are Arabs and all my daughters are married to Arab *mujahadeen* [warriors of the *jihad*]. I have a spiritual relationship with Mullah Omar. He is a great and brave Muslim of this age. He does not fear anyone but Allah. He is not under any personal relationship or obligation to me. He is only discharging his religious duty.

INTRODUCTION

Why We Fight America

by Sulaiman Abu Ghaith

One of America's most wanted al Qaeda operatives, Sulaiman Abu Ghaith is a former Kuwaiti religious studies teacher who became famous during the Iraqi occupation of his country in 1990, when he gave fiery sermons daring Kuwait to attack Baghdad and the forces of Saddam Hussein. After the liberation of his country during the Persian Gulf War, Abu Ghaith became banned from preaching because his sermons had turned against the Kuwaiti government and other Arab states, including Saudi Arabia. In 2001 his Kuwaiti citizenship was revoked when he appeared on Al-Jazeera, the Arab television network, as the official spokesman for al Qaeda. The following address was posted on an official al-Qaeda Web site from where it was downloaded and verified as authentic by the FBI. The address was later published by the Jerusalem Post *and the Center for Islamic Research and Studies (www.alneta.com) and the Middle East Media Research Institute, a nonprofit organization in Washington, DC.*

WHY WE FIGHT AMERICA

Perhaps the [Islamic] nation is waiting for one al Qaeda man to come out and clear up the many questions that accompany any communiqué, message, or picture [concerning September 11], to know the truth, the motives, and the goals behind the conflict with the Hubal [one of the pre-Islamic Kaaba idols—referring to the United States] of our generation.

Why is the world surprised?! Why were millions of people astounded by what happened to America on September 11? Did the world think that anything else would happen? That something less than this would happen?!

What happened to America is something natural, an expected event for a country that uses terror, arrogant policy, and suppression against the nations and the peoples, and imposes a single method, thought, and way of life, as if the people of the entire world are clerks in its government offices and employed by its commercial companies and institutions.

Anyone who was surprised, and did not expect [the events of September 11,] did not [understand] the nature of man, and the effects of oppression and tyranny on man's emotions and feelings. They thought that oppression begets surrender, that repression begets silence, that tyranny only leaves humiliation. Perhaps they also thought that this [oppressive] atmosphere is sufficient to kill man's virility, shatter his will, and uproot his honor. These people erred twice: once when they ignored [the consequences of] treating man with contempt, and again when they were unaware of man's ability to triumph.

This goes for every man—let alone when the man in question is of those who believe in Allah, in Islam as a religion, and in Muhammad as Prophet and Messenger, and anyone who knows that his religion is unwilling to allow him to be inferior and refuses to allow him to be humiliated.

THE ENTIRE EARTH MUST BE SUBJECTED TO ISLAM

How can [he] possibly [accept humiliation and inferiority] when he knows that his nation was created to stand at the center of leadership, at the center of hegemony and rule, at the center of ability and sacrifice? How can [he] possibly [accept humiliation and inferiority] when he knows that the [divine] rule is that the entire earth must be subject to the religion of Allah—not to the East, not to the West—to no ideology and to no path except for the path of Allah?

As long as this Muslim knows and believes in these facts, he will not—even for a single moment—stop striving to achieve it, even if it costs him his soul . . . his time, his property, and his son, as it is said, Say [to the believers]: If your fathers and your sons and your brethren and your wives and your kinsfolk and the worth you have acquired and the trade, the dullness of which you apprehend, and the dwellings that you fancy are dearer to you than Allah and His Messenger, and striving in His cause, then wait until Allah issues His judgment. Allah guides not the disobedient people.

THE BLOW AGAINST THE UNITED STATES WILL COME FROM WHERE IT IS LEAST EXPECTED

The [premises] on which we base ourselves as an organization, and on which we base our operations and our method of action, are practical and realistic. . . . They are also scientific and [in accordance with] Islamic religious law, and they give us confidence and certainty. . . . In writing them and in [publicly] revealing them, I do not intend to be apologetic for what was done; I lay [these arguments] before you so as to emphasize that we are continuing with our blows against the Americans and the Jews, and with attacking them, both people and installations [so as to stress] that what awaits the Americans will not, Allah willing, be less than what has already happened to them. America must prepare itself; it must go on maximum alert; . . . because, Allah willing, the blow will come from where they least expect it.

America is the head of heresy in our modern world, and it leads an infidel democratic regime that is based upon separation of religion and state and on ruling the people by the people via legislating laws that contradict the way of Allah and permit what Allah has prohibited. This compels the other countries to act in accordance with the same laws in the same ways . . . and punishes any country [that rebels against these laws] by besieging it, and then by boycotting it. By so doing, [America] seeks to impose on the world a religion that is not Allah's.

America, with the collaboration of the Jews, is the leader of corruption and the breakdown [of values], whether moral, ideological, political, or economic corruption. It disseminates abomination and licentiousness among the people via the cheap media and the vile curricula.

America is the reason for all oppression, injustice, licentiousness, or suppression that is the Muslims' lot. It stands behind all the disasters that were caused and are still being caused to the Muslims; it is immersed in the blood of Muslims and cannot hide this.

For fifty years in Palestine, the Jews—with the blessing and support of the Americans—carried out abominations of murder, suppression, abuse, and exile. . . . The Jews exiled nearly 5 million Palestinians and killed nearly 260,000. They wounded nearly 180,000, and crippled nearly 160,000.

Due to the American bombings and siege of Iraq, more than 1,200,000 Muslims were killed in the past decade. Due to the siege, over a million children are killed [annually]—that is 83,333 children on average per month, 2,777 children on average per day. Five thousand Iraqis were killed in one day in the Al-Amiriya shelter alone.

In its war against the Taliban and al Qaeda in Afghanistan, America has killed 12,000 Afghan civilians and 350 Arab *jihad* fighters, among them women and children. It annihilated entire families from among the Arab *jihad* fighters while they were in their cars, when the American Air Force bombed [them] with helicopters and antitank missiles, until nothing remained of some of them except scattered body parts.

In Somalia, America killed 13,000 Somalis and [its soldiers] carried out acts of abomination on [Somali] boys and women.

Muslims Have Suffered from America's Standing with Christians

America's standing with the Christians of the world against the Muslims has stripped the camouflage from its face. Much can be said about this regarding Sudan, the Philippines, Indonesia, Kashmir, Macedonia, Bosnia, and other tragedies. America's siege on the Islamic countries as punishment for their rebellion against its laws has transgressed all limits, and Muslims have suffered economic losses that outstrip the imagination.

After all this, is it forbidden for a victim to escape when he is tied and brought to the slaughterhouse?!! Is he not entitled, while he is being slaughtered, to stamp his feet?!!

After all this, some [Arab regimes] shed crocodile tears for what happened to the country of heresy [America], and tried to exonerate Islam from what happened to [America] and asked the country of heresy to treat the Muslims sensitively and gently, and sent messengers and broadcasters to the Jihad fighters with a request to stop fighting Hubal [i.e., meaning the United States]. Do they really think we would do this?

No, by Allah. They [the Arab regimes] have turned their back on us and we have turned our back on them. . . . We would have no honor if we did not avenge the blood of our brothers in Palestine, in Iraq, in Afghanistan, and everywhere.

The banner is being waved openly, and now there is "only a trench of belief" and "a trench of heresy."

The Islamic Justification for al Qaeda's Jihad against the United States

The religious arguments on which we base ourselves in our *jihad* against the Americans—the explanations that inspire us with confidence in the triumph of our religion, our belief, and our faith—are many, and this is not the place to enumerate them, as they are included in the books of the sages.

No one disagrees with these explanations, except he who lives [in] fear . . . he who asks for shelter, thinking that he has distanced himself from evil . . . or he who kneels as a doorman before the doors of the tyrants to gain a position, advancement, or a gift!!

These people have not, Allah be praised, dissuaded us, not even for a single day, from continuing in our path, from our *jihad*, and from our mission. Allah willing, they will not prevent us [in the future].

In this article I will present one explanation that suffices [to wage] *jihad* against the Americans, the Jews, and anyone who has gone in their path.

Allah said, "He who attacked you, attack him as he attacked you," and also, "The reward of evil is a similar evil," and also, "When you are punished, punish as you have been punished."

The words of the sages on these verses are clear: Ibn Taimiyyah [in his book] *Al-Ikhtiyarat Wa-Al-Fatawi*; Ibn al-Qayim in *I'lam Al-Muqi'in* and in *Al-Hashiya*; al-Qurtubi in his *Tafsir*; al-Nawawi in *Al-Muhazab*; al-Shukani in *Nayl Al-Awtar*; and others, may Allah's mercy be upon them.

Anyone who peruses these sources reaches a single conclusion: The sages have agreed that the reciprocal punishment to which the verses referred is not limited to a specific instance. It is a valid rule for punishments for infidels, for the licentious Muslims, and for the oppressors.

ISLAMIC LAW ALLOWS RECIPROCATION AGAINST THE UNITED STATES

If by religious law it is permitted to punish a Muslim [for the crime he committed]—it is all the more permitted to punish a *Harbi* infidel [i.e., he who belongs to *dar al-Harb*, "the domain of disbelief"] in the same way he treated the Muslim.

According to the numbers I noted in the previous section of the lives lost from among the Muslims because of the Americans, directly or indirectly, we still are at the beginning of the way. The

Americans have still not tasted from our hands what we have tasted from theirs. The [number of] killed in the World Trade Center and the Pentagon were no more than fair exchange for the ones killed in the Al-Amiriya shelter in Iraq, and are but a tiny part of the exchange for those killed in Palestine, Somalia, Sudan, the Philippines, Bosnia, Kashmir, Chechnya, and Afghanistan.

WE HAVE THE RIGHT TO KILL FOUR MILLION AMERICANS

We have not reached parity with them. We have the right to kill 4 million Americans—2 million of them children—and to exile twice as many and wound and cripple hundreds of thousands. Furthermore, it is our right to fight them with chemical and biological weapons, so as to afflict them with the fatal maladies that have afflicted the Muslims because of the [Americans'] chemical and biological weapons.

America knows only the language of force. This is the only way to stop it and make it take its hands off the Muslims and their affairs. America does not know the language of dialogue!! Or the language of peaceful coexistence!! America is kept at bay by blood.

June 2002

PART ONE

THE ISLAMIC MAFIA

The terrible calamity!
What is the terrible calamity?
And what will make you comprehend what the terrible calamity is?
The day on which men shall be as scattered moths,
And the mountains shall be as loosened wool.
Then as for him whose measure of good deeds is heavy,
He shall live a pleasant life.
And as for him whose measure of good deeds is light,
His abode shall be the abyss.
And what will make you know what it is?
A burning fire.

—"The Terrible Calamity," Koran 101:1–11

CHAPTER ONE

The Young Lion and the Dream
of the American Hiroshima

Allah has ordered us to glorify the truth and to defend Muslim land, especially the Arab peninsula . . . against the unbelievers. After World War II, the Americans grew more unfair and more oppressive towards people in general and Muslims in particular. . . . The Americans started it and retaliation and punishment should be carried out following the principle of reciprocity, especially when women and children are involved. Through history, America has not been known to differentiate between the military and the civilians or between men and women or adults and children. Those who threw atomic bombs and used the weapons of mass destruction against Nagasaki and Hiroshima were the Americans. Can the bombs differentiate between military and women and infants and children? America has no religion that can deter her from exterminating whole peoples. Your position against Muslims in Palestine is despicable and disgraceful. America has no shame. . . . We believe that the worst thieves in the world today and the worst terrorists are the Americans. Nothing could stop you except perhaps retaliation in kind. We do not have to differentiate between military or civilian. As far as we are concerned, they are all targets, and this is what the fatwa says.
—Osama bin Laden, interview with John Miller, 1998

Humanity today is living in a large brothel! One has only to glance at its press, films, fashion shows, beauty contests, ballrooms, wine bars and broadcasting stations! Or observe its mad lust for naked flesh, provocative pictures, and sick, suggestive statements in literature, the arts, and mass media! And add to all this the system of usury which fuels man's voracity for money and engenders vile methods for its accumulation and investment, in addition to fraud, trickery, and blackmail dressed up in the garb of law.

—Sayyid Qutb, *Milestones,* 1965

Islam is a revolutionary ideology. It seeks to alter the social order of the entire world and rebuild it in conformity with its own tenets and ideals. . . . "Muslims" is the title of that "International Revolutionary Party" organized by Islam to carry out its revolutionary program. "Jihad" refers to the revolutionary struggle to achieve this objective.

—Sayyid Abul Ala Maududi, *Towards Understanding Islam,* 1996

In May 1996, Osama bin Laden, the great emir of the holy war, was broke and living in Afghanistan with his four wives, fifteen children, and tribe of followers. He had been deported from Sudan at the insistence of the United States and arrived in Jalalabad on board a Hercules C-130 cargo plane with his family and 140 al Qaeda associates.[1] Although the US intelligence community was fully aware of bin Laden's financial condition, the FBI and the CIA reported that bin Laden's net worth remained in excess of $250 million. This bit of disinformation was passed on to the press and became published as an established fact in major newspapers throughout the world.[2]

Such disinformation was purposeful. It helped to depict bin Laden as a solitary figure, a millionaire *mujahed* [warrior in the *jihad*] who had declared war on America—a war he intended to wage with his own financial resources, not drug money and contributions from millions of Muslims throughout the world.

The truth—as in most matters regarding Osama bin Laden, al Qaeda, and the war on terror—was to the contrary. Proof of bin Laden's real financial condition upon his arrival in Afghanistan came to light in 2001 from testimony of al Qaeda operatives during the proceedings of *The United States v. Osama bin Laden et alia* within a federal courthouse in lower Manhattan. L'Houssaine Kherchtou, a key operative, testified that bin Laden's financial resources had been drained in Sudan, where he had invested a vast sum of money on headquarters for his al Qaeda associates, three large training cells, a construction company called Al-Hajira, a farm that covered thousands of acres in the central province of Gezira, an office building on McNumir Street in Khartoum, a house for his wives and children in the Riyadh district, a farm outside of town, and an Islamic bank—all of these assets he would become forced to relinquish when he became an exile.[3] By 1995, Kherchtou said, bin Laden had been forced to cut the salaries and expenses of all group members "to the bone," and by 1996 the emir could not afford to renew the license for the pilot of his plane because money had become "too tight."[4]

Bin Laden's financial woes were further exacerbated by the millions he spent in his quest for the weapons of mass destruction that were necessary to mount the great *jihad* against the United States. Some insight into this spending was provided by Jamal Ahmed al-Fadl, another al Qaeda operative, during the 2001 trial. Al-Fadl said that he had been singled out by the al Qaeda chieftain to negotiate a deal for the purchase of enriched uranium from two black market agents in Khartoum. One agent purportedly was Salah Abdel al-Mobruk, a lieutenant colonel in the Sudanese army and former government minister; the other was a merchant named Basheer.

The first meeting took place in an office building on Jambouria Street. Al-Fadl was informed that the asking price for the uranium was $1.5 million per kilo, plus payments of commissions for al-Mobruk and Basheer. The two agents said that the money had to be paid outside of Sudan. Al-Fadl relayed the terms to one of bin Laden's top lieutenants, who said that the terms were acceptable as long as the uranium was established to be weapons grade. A second meeting was held between al-Fadl and the two agents in a small house within the village of Bait Al-Mal, north of Khartoum. Here, al-Fadl was shown a small cylinder between two and three feet tall with specification engravings which indicated that it was of South African origin.[5]

After making arrangements for a test of the uranium that was to be conducted with machinery and nuclear technicians from Nairobi, al-Fadl was informed by al Qaeda officials that his services were no longer required. He was paid $10,000 for putting together the deal, a sum that al-Fadl considered paltry. The test was eventually conducted within Hilat Koko in northern Cyprus, but al-Fadl was informed of neither the results nor the final outcome of the negotiations.[6]

In the CIA's bin Laden file (codenamed "Alec") remains verification that the uranium in question had been stolen from Valindaba, the nuclear manufacturing facility near Pretoria in South Africa. Under the apartheid regime, Valindaba had produced weapons-grade enriched uranium—enough to produce twenty nuclear bombs. Each bomb would have an explosive yield of fifteen to twenty kilotons of TNT—far more destructive than the atomic bombs that were dropped on Hiroshima and Nagasaki. In the Alec file, one can find a bank draft of $1.5 million drawn by bin Laden from a Bank of Budapest account to pay for the sample kilo. However, the file provides no proof that the sale was consummated.[7]

Sudan in 1993 was an ideal place for bin Laden to begin his search for weapons of mass destruction. The nuclear black market was booming, and Sudan had become a pivotal transshipping point for weapons and materials, including enriched uranium. These goods flowed from Russia to Germany and from Germany to Sudan. From Sudan, the flow continued to such places as Iran, Libya, and Pakistan.[8] The key al Qaeda agent in charge of the procurement of nuclear weapons was Mamdouh Mahmud Salim, an electrical engineer from Iraq. Salim, who combed the world for loose nukes, was eventually arrested in Germany in 1998, while attempting to buy several kilos of uranium for bin Laden, and extradited for trial in the United States.[9]

How much nuclear material Salim and other al Qaeda operatives were able to procure remains unknown. But the efforts must have resulted in some success since, as the CIA would later reveal, bin Laden established a laboratory in Khartoum in 1993 and hired a physicist from a Middle Eastern country to work on his project to create nuclear weapons that would be capable of killing millions of

Americans.[10] Bin Laden made no secret of his intent. "It is the duty of Muslims to possess these weapons," he said on more than one occasion.[11]

After the bombing of the World Trade Center on February 23, 1993, the United States began to pressure the Sudanese government for bin Laden's expulsion with threats of providing military assistance to Sudan's hostile neighbors: Uganda, Eritrea, and Ethiopia.[12] Additional pressure came from Egypt after al Qaeda's botched attempt to assassinate President Hosni Mubarak on June 26, 1995, and from Saudi Arabia after the terrorist group's bombing of the training center of the Saudi National Guard on November 13, 1995. Osama had become too hot for even Sudanese president Omar al-Bashir, who had sheltered Carlos the Jackal (the mastermind of terrorist bombings, kidnappings, and hijackings throughout the 1970s and early 1980s), to handle.

Backstage diplomatic maneuverings took place to bring about bin Laden's expulsion.[13] Prince Turki bin Faisal, the head of the Saudi intelligence service, later confided to a member of the press: "Sudan offered him [bin Laden] first to the U.S. who turned the offer down as they did not consider they had enough to indict him in an American court of law. So the Sudanese came to us with a proposition that we take Osama back but not try him in our courts. We turned them down and Crown Prince Abdullah told President al-Bashir, 'No one is above the law.'"[14]

By bin Laden's own admission, he had lost more than $150 million in Sudanese ventures.[15] The Saudi government had revoked his citizenship and had frozen his assets.[16] He became a man without a country.

Arriving without ceremony in Afghanistan, Osama bin Laden, the son of Muhammad bin Laden, seemed to have come to the end of his tether. He had been raised in the company of princes and kings. He was a university graduate with engineering and business skills and influential contacts throughout the world. Had he followed the path forged by his father, he could have been a respected engineer and building contractor, a pious member of the *ummah* (community of believers) in Saudi Arabia, and a billionaire in his own right.

Muhammad bin Laden, Osama's father, had left his native village of Al-Ribat in central Yemen in 1925 and settled in the Hejaz Province of Saudi Arabia. While working as a bricklayer on the new imperial palace, Muhammad gained the attention of Ibn Said, who recognized him as a gifted craftsman. He also displayed his ability as a provider of discreet services, such as the laundering of payments to their various causes.

Overnight, the illiterate bricklayer went from rags to riches. Muhammad received the contract to build the first royal palace in the Saudi port city of Jeddah, a contract for the rebuilding and refurbishing of the sacred mosques in Mecca and Medina, and a contract to build a road from Mecca to Taif in the Asir Mountains, where the royal families came to construct their country retreats. Osama once told the press: "It is no secret that my father was responsible for the infrastructure of Saudi Arabia. God blessed him and bestowed on him an honor that no other contractor has known. He built the Holy Mecca Mosque where the Holy Kaabah is located and he built the Holy Mosque in Medina for our Prophet."[17]

This was only the beginning of prosperity for the bin Laden family. The oil boom of the 1970s transformed the company into one of the largest firms in the Middle East. The Bin Laden Corporation received contracts for the building of tens of thousands of housing units, for the construction of roads, such as the major highway between the two holy cities, and for the development of farm lands—replete with irrigation systems—throughout the desert.

Osama, "young lion" in Arabic, was born in 1957 and raised in a household of devout Wahhabists, and he, like his fifty brothers and sisters, was educated in Wahhabist schools. The Wahhabists follow the radical Sunni Islamic tradition espoused by Muhammad ibn Abdul Wahhab (1703–1791), who called for an uncompromising interpretation of the Koran—an interpretation that demands the institution of *shariah* in all Muslim lands: no use of prayer beads, no other name in prayer except Allah, no smoking of tobacco, no drinking of intoxicants, no abusive language, and no ornamentation in places of worship. Recalling his childhood, Osama said: "Every grown-up Muslim hates America, Christians, and Jews. It is part of our belief and our religion. Ever since I was a boy, I have been harboring feelings of hatred toward America."[18]

Alia, Osama's tiny Syrian mother, was one of the least favorites of his father's wives.[19] Muhammad's other three wives were Saudis. The fourth wife, the last allowed by Muslim tradition, represented a movable object, frequently finding herself divorced to make way for a new favorite—a fate that befell Alia. The practice allowed Muhammad to accumulate a total of twenty-one wives during his long and fruitful career.[20]

The elder bin Laden, despite his penchant for dropping and adding the fourth wife, became renowned throughout the kingdom for his religious fervor and his piety. For more than forty years, he awaited the coming of the Mahdi, a Messianic figure foretold by the Haddith, who will lead the Muslim people to great victory over their enemies and bring forth the Day of Islam, when all people—believers and nonbelievers alike—fall in total submission before the throne of Allah. Muhammad even established a charitable fund of $12 million to assist the Mahdi in restoring the grandeur of Islam throughout the world. The pious old man could never have dreamed that his quiet and wide-eyed seventeenth son would come to lay claim to this exalted title. "He was fascinated—obsessed—by religion," said a businessman who had worked with Muhammad for many years. "He loved obscure religious debate, and spent huge sums financing these regular evening meetings called *halqas* where the greatest preachers and religious teachers in the Saudi kingdom would gather to debate theology. I guess it satisfied some philosophical streak within him."[21]

When Muhammad bin Laden died in a plane crash sometime between 1967 and 1973,[22] King Faisal, who reportedly had only wept once before in his life, was said to have shed a second tear over the unfortunate demise of his close friend and favorite builder.[23]

The Bin Laden Corporation now came under the leadership of Muhammad's sons Salem, Bakr, and Ali, who adopted the new name of Bin Laden Brothers for Contracting and Industry and established corporate headquarters in Jeddah. The Bin Laden Brothers came to represent a host of major European companies in the Middle East, including Audi, Volkswagen, and Porsche automobiles, and became the distributor of luxury products, such as the line of goods produced by the Dutch company Pander Projects, and of soft drinks, such as Snapple. It also teamed up with Hunting Survey Ltd., a British company, for the construction of prefabricated buildings. The company also continued to receive contracts for huge construction projects throughout the Saudi kingdom, including a massive highway around Riyadh, a reception hall for the royal palace, a high rise and a luxury hotel in Mecca, and the construction of the terminal and hangars at the Riyadh airport. As it expanded, the Bin Laden Brothers became known as the Saudi Bin Laden Group.[24]

By September 11, 2001, the group counted among its business partners such firms as General Electric, Citicorps, and the Carlyle Group. The latter represented an investment company whose principals included former secretary of defense Frank Carlucci, former secretary of state James Baker, and former British prime minister John Major; and whose advisors included former US president George H. W. Bush.[25]

There were also ties between the Bin Laden Group and other prominent political figures. In 1973 Salem bin Laden incorporated a company called Bin Laden Aviation in Austin, Texas. In 1976 Salem purportedly named James Bath, an aircraft broker from Houston, as his American agent and trustee.[26] Bath, on behalf of bin Laden, was said to have purchased several airplanes from Air America, a company in northeastern Pennsylvania. Air America, which was headed by Rik Luytjes, was a shady outfit that served to transport over $2 billion worth of cocaine from Colombia to the United States.[27] Luytjes became friendly with many Republican luminaries, including William Warren Scranton, the former governor of Pennsylvania and UN ambassador, and Joseph M. McDade, a powerful member of the House of Representatives, who chaired the Appropriations Committee.[28]

Soon after being appointed as Salem bin Laden's trustee, Bath became a partner with George W. Bush in an oil company called

Arbusto Energy. The investment money came from Salem bin Laden. Later, when questioned by the press, George W. Bush denied ever knowing Bath, although they had served together in the Texas Air National Guard. Finally, the future president admitted that Bath possessed a stake in Arbusto (which means "bush" in Spanish) and that he was aware that Bath represented Saudi interests.[29]

In 1988 Rik Luytes was arrested, and the story of Air America captured national headlines. Days after the story broke, on May 29, 1988, Salem bin Laden was killed in a mysterious plane crash near San Antonio, when his ultralight aircraft crashed into power lines immediately after takeoff. The eldest son of Muhammad bin Laden had been an experienced pilot with more than fifteen thousand hours of flight experience.[30] The aircraft Salem was flying was one of the planes he had purchased from Air America.

When his father, the nucleus of the bin Laden family also died in a plane crash, though in Saudi Arabia, Osama inherited a fortune estimated at $80 million—$30 million in cash and $50 million tied up in the company business—enough to grant him a place of prominence within the Saudi business community.[31]

Although Osama likes to present himself as the model of a believing Muslim who shared his father's fundamentalist beliefs and piety, he made regular trips to Beirut, then the fleshpot of the Arab world, after his graduation from high school in 1973. Within the Lebanese capital, he gained a reputation as a binge drinker with a weakness for prostitutes and bar girls. His older brothers complained to friends and business associates of young Osama's profligate behavior, while reminding themselves and others that he was the "son of a slave."[32]

By 1975, Osama tired of the excesses of youth and took, through an arranged marriage, a young Syrian girl as his first wife. Her name was Najwa Ghanem, and she would bear his first son, Abdullah.[33] The same year, Osama entered King Abdul Aziz University in Jeddah, where he studied economics, agriculture, and civil engineering.

At Jeddah, he came under the influence of two powerful figures. The first was Abdullah Azzam, a teacher of religion and a leader of the Muslim Brotherhood, a group that advocated the imposition of Islamic law (*shariah*) on all Muslim nations and the creation of a pan-Islamic state. The second was Muhammad Qutb, who taught Islamic studies. Muhammad Qutb introduced Osama to the writings of Sayyid Qutb, Muhammad's esteemed brother and the so-called father of modern Islamic fundamentalism. Sayyid Qutb, who had been hanged in a Cairo jail for sedition in 1965, had maintained that Christians and Jews remained headed for hell and that Muslim leaders (such as President Nasser of Egypt and King Saud of Saudi Arabia) who did not abide by an intransigent interpretation of the teachings of the Koran were apostates, who must be resisted and overthrown.[34]

While studying at the university, Osama was also attending business meetings at the family business. He was clean-shaven and soft-spoken and always appeared at the firm in a well-tailored Western suit and tie.[35]

In 1979 eighty-five thousand Soviet troops invaded Afghanistan to prop up the communist government of Afghan president Noor Takaki. Religious teachers and clerics throughout Saudi Arabia began calling upon young men to take up arms in a *jihad* to liberate their Muslim brothers. Osama became one of the first to answer the call. By so doing, Osama believed that he was abiding by the wishes of his dead father: "My father was very keen that one of his sons should fight against the enemies of Islam. So I am the one son who is acting according to the wishes of his father."[36]

On December 26, 1979, Osama boarded a plane and headed off for a training camp in Pakistan. He took along his four wives (a Saudi, in addition to Najwa, a Syrian, and a Philippine) and some fifteen children.[37] Milt Beardon, the CIA station chief in Peshawar, described the Arab recruits who arrived in the Pakistani city as follows: "Some were genuine, on missions of humanitarian value, while others were adventure-seekers looking for paths of glory, and still others were psychopaths."[38]

Osama, at the age of twenty-three, was incredibly tall at six feet five and incredibly thin, weighing less than 180 pounds. He appeared to be ungainly, but he was quite athletic and highly skilled as a horseman, mountain climber, and soccer player.

During the next ten years, Osama engaged in armed conflict (including hand-to-hand combat against members of the Red Army in Jaji and Shaban) and helped build tunnels, roads, and bunkers using heavy construction equipment that he had purchased with his inheritance.[39] Should you wish to take your life in hand, you can still visit the huge tunnels that he blasted into the Zazi Mountains of the Bahktiar Province for use as guerilla hospitals and ammunition dumps.[40]

In 1984 Abdullah Azzam, Osama's former university professor, set up Mekhtab al-Khadamat ("the Office of Services"), a recruiting center for the *jihad* in Peshawar. When bin Laden returned from the front, he would stay with Azzam and the new recruits and sleep on the floor. Azzam would like to tell the recruits, "This man has everything in his own country. You see, he lives with all the poor people in this room."[41] The two became fast friends. Azzam traveled throughout the Middle East, the United Kingdom, and the United States to raise money and recruits for the holy war, while Osama provided financial support, handled military matters, and brought to Afghanistan experts from around the world in guerrilla warfare, sabotage, and covert operations.

The Soviet-Afghan War set the stage for the last standoff between the Soviet Union and the United States. To tip the scales against the Soviets, President Jimmy Carter provided the *mujahadeen* with $30 million in covert aid. This amount increased under the Reagan administration, and so did the carnage and the number of refugees. By 1985 the Afghan rebels were receiving $250 million a year in covert assistance to battle the by now 115,000 Soviet troops. This figure was double 1984's amount. The annual amount received by the guerrillas reached $700 million by 1988. By this time, the CIA was even shipping Tennessee mules to Afghanistan to carry all the

weapons in the hills.[42] Even now, in the wake of Operation
Enduring Freedom, Afghanistan remains littered with the abandoned one-way shipping containers that were used to bring all the
weaponry into the country. In all, the Soviets said that 14,453 Red
Army solders had been killed in the conflict. The real number,
according to US intelligence sources, may be closer to 35,000.[43]

Toward the end of the war, bin Laden and Azzam met with members
of the Egyptian Islamic Jihad, the organization responsible for the
assassination of Egyptian president Anwar Sadat. The leader of the
group was Ayman al-Zawahiri, a physician from Cairo, who had
arrived in Afghanistan in 1980. With his poor vision and lack of military training, al-Zawahiri proved to be of little use in combat, but
he was extremely valuable in providing medical care and treatment
for the troops.[44] He courted bin Laden with the hope of obtaining
his financial support for a few of his pet projects, including the
assassinations of President Muhammad Zia ul-Haq of Pakistan and
President Hosni Mubarak of Egypt, who, in his opinion, had become apostates.[45]

Bin Laden was impressed with al-Zawahiri, whom he called "the
Doctor," because of his education, his erudition, and his hatred of
Soviets, Jews, and Christians.[46]

As a gesture of his friendship, the "Doctor" loaned two of his
men to bin Laden: Muhammad Atef and Abu Ubaidah al-Banshiri.
Atef later became al Qaeda's operational planner. He was killed in a
US air strike on Afghanistan in November 2001. Al-Banshiri came to
serve as al Qaeda's second in command of military operations. He
drowned in a ferry boat accident on Lake Victoria in May 1996.[47]

Subsequent meetings were held until al-Zawahiri and the Egyptians became regular guests at Mekhtab al-Khadamat. Bin Laden
gradually drifted away from his old friend Azzam and toward his
new friends. The office transformed into a battleground for ideological differences. Azzam argued that the leaders of the *mujahadeen*
should concentrate their efforts on the creation of a model Islamic

state in Afghanistan that could capture the support of Muslims throughout the world and, in turn, wage war against Israel. Al-Zawahiri disagreed, as did bin Laden, arguing that a war should be fought on two fronts: the first against *kafir* countries, including the United States; and the second against apostate Muslim leaders, such as Hosni Mubarak of Egypt and Muammar Gadhafi of Libya.[48] Soon Azzam no longer had a say in the operations of the Office of Services that he had labored to create, and bin Laden returned to Afghanistan to create a training camp for Arab commandos that he called Al Mashah, "the lion's den."[49]

On November 24, 1989, Azzam and his two sons, Ibrahim and Muhammad, were killed by a car bomb containing at least twenty kilos of TNT in Peshawar as they drove to a mosque for Friday prayer services.[50] The bomb was activated by remote control.[51] The murders were never solved.

Shortly before the Soviet withdrawal in 1989, bin Laden and the Egyptians met at Al Mashah to create a headquarters to carry out their plans of continuing the *jihad* throughout the world. The guiding principle for the establishment of al Qaeda ("the Base"), no less than Marxism, was a utopian vision—the dream of a new world order that had been expressed by such Islamic scholars as Ibn Tamiyya, Rashid Rida, and Sayyid Qutb. It called for the creation of an empire—a pan-Islamic state—that would unite the world's one billion Muslims under a single ruler and a single system of secular and religious law, the *shariah*. The instrument that would serve to create this universal *caliphate* would be al Qaeda, and the means to accomplish this religious and political objective would be *jihad*, a holy war that would rage until all of creation fell before the throne of Allah.

The task, as Qutb had pointed out, entailed "a full revolt against human rulership in all its shapes and forms, systems and arrangements." It meant "destroying the kingdom of man to establish the kingdom of heaven on earth."[52]

The war in Afghanistan had taught bin Laden and al-Zawahiri

that this goal was attainable. After all, the guerrilla army of Muslims had brought the Soviet Union, one of the world's superpowers, to its knees. This triumph was a result of their steadfast faith in Allah and the words of the Prophet, not of the billions in money and matériel that had been provided by the United States. "The lesson here is that *jihad* is a duty for the nation," bin Laden said. "We believe that those who waged *jihad* in Afghanistan performed a great duty. They managed with their limited resources of RPGs, antitank mines, and Kalashnikovs to defeat the biggest legend known to mankind, to destroy the biggest war machine, and to remove from our minds the so-called big powers."[53]

To bring forth the new *caliphate*, the apostate leaders of pro-Western Islamic countries (Saudi Arabia, Jordan, and Egypt) would have to be replaced with righteous rulers, and the Western powers that propped up these apostate states would have to be destroyed. Chief among such powers was the United States of America. If the United States was destroyed, these puppet regimes in the Middle East would collapse and the new *caliphate* would emerge from the rubble.

How could the guerrilla band in Afghanistan hope to bring about the destruction of the mightiest nation in the world? The answer, bin Laden and al-Zawahiri realized, lay in the acquisition of weapons of mass destruction, including, most particularly, nuclear weapons. The United States had unleashed such weapons against the civilian population of Japan at the close of World War II. Therefore, in accordance with the Islamic teaching of parity, they saw it was only fitting that such weapons be used against the American people.

According to testimony provided by Jamal Ahmed al-Fadl and other informants, al Qaeda became structured in the following manner:

- **Emir**—Osama bin Laden
- **Chief Counsel**—Ayman al-Zawahiri
- **The *Shura* or Consultation Council**—the group of seasoned leaders of the *jihad*, who must approve major decisions such

as terrorist attacks and the issuance of *fatwas* (edicts). The members of the *shura* (Majilis al-Shura) play leadership roles in al Qaeda's major committees that consist of lower-ranking members of the organization.

The Committees

- **The Military Committee** oversees recruitment, training, and the purchase of weapons.
- **The Islamic Study Committee** makes rulings on religious law and trains all recruits in the teachings of the Koran and the Hadith (the tradition of the Prophet)
- **The Finance Committee** oversees investments, corporate holdings, and tribute (monetary payments) exacted from Middle Eastern countries in exchange for bin Laden's promise not to establish al Qaeda cells within their borders. This latter source of income is akin to the *Danegeld* that was paid by Anglo-Saxon kings to Viking princes or protection money that was provided to the Mafia.
- **The Cells**—the selected sites of operatives throughout the world. Each cell engages in missions independent of the others, and its members, activities, and locations are kept secret from the other cells.
- **Headquarters**—Al Qaeda's first center of operation was on the outskirts of the city of Peshawar in Pakistan. The center, according to the latest intelligence, is now in the valley of Dir in the North-West Frontier Province of Pakistan, less than eighty miles from its original base.[54]

Al Qaeda, al-Fadl maintained, also employed the services of Abu Muaz el-Masrry, an interpreter of dreams. He attended all meetings of the *Shura* and met with bin Laden on a daily basis.[55]

A list of al Qaeda members, many of whom served with bin Laden in Afghanistan, and their whereabouts, as of this writing, is as follows:

- **Osama bin Laden**—at large
- **Ayman al-Zawahiri**—at large
- **Saif al-Adel** (aka Saif Adel Makkawi), (security chief, Consultation Council and Military Committee)—at large
- **Muhammed Atef** (Operations Chief)—killed
- **Khalid Shaikh Mohammed** (operations planner)—captured
- **Mafouz Ould Walid** (counselor)—unknown, presumed killed
- **Abd Rahim Nashiri** (Persian Gulf Operational Coordinator)—captured
- **Abu Musab al-Zarqawi** (aka Ahmad Fadeel Khalayleh), (Operational Planner)—at large
- **Zayn Abidin Muhammed Hussein** (aka Abu Zubaydah), (Operational Planner)—captured
- **Tawfiq Attash** (aka Khallad), (Pakistan Coordinator)—at large
- **Zaid Khayr** (operational commander)—at large
- **Saad bin Laden** (son of Osama bin Laden)—at large
- **Abu Ali Harithi** (aka Abu Ali), (USS *Cole* attack planner)—killed
- **Abdullah Ahmed Abdullah** (chief financial officer, Consultation Council and Religious/Fatwa Committee)—at large
- **Ridwuan Isamuddin** (aka Hambali), (Southeast Asia Coordinator)—at large
- **Ramzi Binalshibh** (9/11 Coordinator in Hamburg)—captured
- **Ab Haili** (operational planner, recruiter)—captured
- **Midhat Mursi** (aka Abu Khabab), (chemical weapons specialist)—unknown, perhaps captured
- **Ibn Shaykh Liby** (trainer—training camp commander)—captured
- **Sulaiman Abu Ghaith** (aka Gaith Abu Yusuf), (official spokesman)—at large
- **Fazul Abdullah** (operational coordinator)—unknown
- **Saleh Abdullah** (operational coordinator)—unknown
- **Shaykh Said Bamusa** (financial aide)—unknown

- **Abu Ubaidah al-Banshiri** (cofounder of al Qaeda, head of Military Committee)—deceased
- **Tariq Anwar Fathy** (al Qaeda Aide—Egyptian Islamic Jihad)—killed
- **Abd Hadi al-Iraqi** (trainer—Camp Commander)—detained
- **Abu Jafar Jaziri** (operational coordinator—logistics)—presumed killed
- **Anas al-Liby** (Consultation Council)—unknown
- **Abu Kahlid al-Masri** (senior aide)—unknown
- **Muhammad Sala Rahman** (al Qaeda aide—Egyptian Islamic Jihad leader)—killed
- **Assadullah Rahman** (aide—son of blind Egyptian Muslim cleric involved in 1993 WTC bombing)—killed
- **Omar Mahmoud Othman Omar** (aka Abu Qutadah), (Fatwa Committee)—unknown
- **Abdul Rahim-Riyadh** (operational coordinator—facilitator)—unknown
- **Mamdouh Mahmud Salim** (cofounder of al Qaeda, Camp Manager)—incarcerated in United States
- **Abu Ubaida** (trainer)—killed
- **Abu Bashir al-Yemeni** (trainer—camp commander)—unknown
- **Abu Saleh al-Yemeni** (operational coordinator—facilitator)—killed[56]

After the Soviet withdrawal from Afghanistan, bin Laden returned to Saudi Arabia, where he was greeted as a hero and received invitations to speak at mosques, schools, and homes. More than 250,000 cassettes of his speeches were produced and sold in shops and marketplaces. In these taped speeches, bin Laden decried the evidence of Western cultural imperialism and moral degradation that appeared throughout the kingdom: women disobeying Islamic law in their dress and demeanor, men flaunting their business exploits at social gatherings and drinking alcohol. "When we buy American goods, we are accomplices in the murder of Palestinians," he railed on one of the tapes. "American companies make millions in the Arab world with which they pay taxes to their government. The United States uses that money to send $3 billion a year to Israel, which it uses to kill Palestinians."[57]

On August 2, 1990, Saddam Hussein, president of Iraq, ordered the invasion of Kuwait, forcing the al-Sabagh royal family to flee their palaces and seek refuge in Saudi Arabia. The Saudi royal family reacted with great alarm to the invasion, fearing that Hussein might soon turn his attention to their country, which possessed the largest reserves of crude oil in the world.[58] To ward off this threat, King Fahd sent out a call of alarm to the United States.

Such a call for protection to a *kafir* country was unprecedented. The prophet Muhammad had commanded that two religions must not exist in Arabia. The Muslims, from time immemorial, understood this injunction to mean that nonbelievers should not be permitted to live anywhere within the Arabian Peninsula. For this reason, King Fahd and the royal family prevailed upon Abdul Aziz bin Baz, the Grand Mufti of the Saudi Arabia, to issue a religious decree that would allow this transgression of holy law.[59]

Within a matter of days, giant US C-130 transporters landed at the airbase in Dhahran to discharge hundreds of tanks, trucks, artillery transport vehicles, and jeeps. Satellite dishes sprouted from the rooftops of hotels, office buildings, and apartment complexes. And long lines of US troops, including women in military uniform, walked along the highway between the holy cities of Mecca and Medina.

What could be more of a greater effrontery to the *ummah*, the pious community of believers! Bin Laden was seething with rage and indignation. He wrote: "The Arabian Peninsula has never—since Allah made it flat, created its desert, and encircled it with seas—been stormed by any forces like the Crusader armies now spreading in it like locusts, consuming its riches and destroying its plantations. All of this is happening at a time when nations are attacking Muslims like people fighting over a plate of food."[60] He began to circulate pamphlets and deliver speeches to decry the

"American invasion." Sheikh Safar bin Abd al-Rahman al-Hamali, Sheikh Salman bin Fahd al-Awda, and other clerics and scholars who joined in the protest were rounded up and cast in prison.[61] Bin Laden came to believe that the Saudi rulers were false Muslims (*jahiliyya*) who had to be overthrown and replaced by a body of true believers. In the fall of 1991, Saudi security police uncovered evidence that linked bin Laden to an attempt to smuggle weapons into the kingdom from Yemen as part of a plot to destabilize the government.[62] The Saudi government, despite the pleas of Osama's friends and family, ordered his immediate expulsion from the kingdom.[63]

Osama moved to Sudan, which proved to be an ideal country to rebuild his terrorist organization. Sudan had abandoned visa requirements for Arabs and was openly encouraging Muslim militants from around the world to live within the safety of its borders.[64] By the end of 1991, somewhere between one and two thousand al Qaeda members had settled in Sudan to set up scores of training camps throughout the country, the main one being a twenty-acre site near Soba, seven miles south of Khartoum, along the Blue Nile.[65]

Now, in 1996, Osama bin Laden, at the age of thirty-eight, had returned to Afghanistan, which had become a veritable hell on earth. After seventeen years of continuous warfare, the country had been completely devastated. Nearly one and a half million people had been killed and two million more had migrated to other countries. The irrigation systems in the rural areas had been destroyed, along with the infrastructures within the major cities. Much of the farmland could not be cultivated for crops since it had been sown with tens of thousands of landmines. The vast majority of Afghans existed on a level of bare subsistence. The life expectancy stood at forty-six years, and the infant mortality climbed to 147 per 1,000 births, twenty times that of Europe and the United States.[66] Moreover, the country remained in a state of violent political upheaval, with warlords and their marauding armies of brigands and mercenaries combing the land to rape, pillage, and kill.

Bin Laden came under the protection of the Shii Hazara clan. Other groups, such as the Uzbeks of Gen. Abdul Rashid Dostum and the Tajiks of Cdr. Ahmad Shah Massoud, would have gladly turned him over to US officials for a fistful of greenbacks. But the Clinton administration sought neither to secure his arrest nor to curtail his activities.

As civil war waged in Kabal and Kandahar, bin Laden and his al Qaeda associates established a hideout within a high cave of the Hindu Kush Mountains outside the eastern town of Jalalabad. In November, Abdel Bari Atwan, the editor of the London Saudi weekly *Al-Quds Al-Arabi*, visited bin Laden at his hideout and described the living conditions as follows: "It was not comfortable. His [sleeping] quarters were built in an amateurish way with branches of trees. He had hundreds of books, mostly theological treatises. I slept on a bed underneath which were stored many grenades. I slept maybe a half an hour. I saw perhaps twenty to thirty people around him: Egyptians, Saudis, Yemenis, and Afghans. At night it was very cold, fifteen degrees below zero."[67]

Yet bin Laden was more at peace with his new surroundings than he had been at his expansive villa in Sudan or his luxurious palace in Saudi Arabia. The peace came from his belief that his relocation had been providential—a recapitulation of the Hijra, the prophet Muhammad's migration to Medina. The Prophet had fled Mecca when his *ummah* (his community of believers) became imperiled by hostile tribes—just as he had been forced to leave his homeland when his actions offended the royal family and placed the future of al Qaeda, his *ummah,* in jeopardy. The Hijra had been a time for the Prophet to prepare his followers for the final battle and the triumph of Islam. Similarly, his retreat reprensented a time for him to regroup his soldiers and to issue a call to arms to Muslims throughout the world in preparation for the final battle against the Great Satan.

Allah had led him to this place and granted him an assurance of victory.

This assurance lay before his eyes—in the boundless fields of poppies.

CHAPTER TWO

The Good Life among the Taliban

So off he [Osama bin Laden] goes to Afghanistan, which is probably the best move since the Germans put Lenin in a boxcar and sent him to St. Petersburg in 1917. By forcing him to leave, he was out of the place where we might have been able to control him, or at least monitor him more closely, to see what he was up to.

—Milton Beardon,
former CIA station chief in Peshawar, Pakistan

Afghanistan is the only country in the world with a real Islamic system. All Muslims should show loyalty to the Afghan Taliban leader, Mullah Mohammed Omar.

—Osama bin Laden, April 2001

Today your brothers and sons, the sons of the two Holy Places, have started their Jihad in the cause of Allah, to expel the occupying enemy out of the country of the two Holy Places. And there is no doubt you would like to carry out this mission too, in order to establish the greatness of the Ummah and to liberate its occupied sanctities. Nevertheless, it must be obvious to you that, due to the imbalance of power between our armed forces and the enemy forces, a suitable means of fighting must be adopted.

—Osama bin Laden, "Declaration of War against
the Americans Occupying the Land of the Two Holy Places,"
August 23, 1996

One month before bin Laden's arrival in Afghanistan, Mullah Mohammed Omar, the burly, bearded, one-eyed leader of the Taliban, appeared on the rooftop of a mosque in Kabul and wrapped himself in the cloak of the prophet Muhammad, the most sacred relic in Islam, while thousands of his followers cheered, *"Amir-ul Momineen"* ("Commander of the Faithful")—the honorific reference to a *caliph*.[1] By receiving this title, Omar assumed the position of the highest authority of Sunni Muslims this side of paradise and gave substance to the al Qaeda dream of a pan-Islamic state. Small wonder that Osama felt that he was returning to his spiritual home.

Mullah Omar was not only an Islamic cleric and scholar but also a veteran and hero of the ten-year *jihad* against the Soviet Union. He was born in 1959 in Nodeth, near Kandahar. He rose to become a village mullah (a cleric versed in religious law) and ran a *madrassah* (religious school) in Pakistan before joining the *mujahadeen* when the Soviet Union invaded Afghanistan in 1979. He fought for several years until he was struck in his left eye by a piece of shrapnel when a Soviet shell hit a mosque during the battle of Singesar. Knowing that his eye was lost and that it could become infected, he allegedly ripped the eyeball from its socket, wiped his bloody hand on the wall of a mosque, and returned to take his part in the engagement with his trusty Kalishnikov submachine gun. Visitors to the mosque in Singesar can still see his bloody fingerprint on the outside wall.[2]

At the end of the war, as Afghanistan fell into a chaotic state of violence and lawlessness, Omar returned to the same little village of Singesar, where he devoted his time and attention to meditation on the *surahs* of the Koran. This idyllic period of thought and contemplation was disrupted when the prophet Muhammad appeared to him in a dream and ordered him to put an end to the reign of terror of the warlords throughout the country.[3] The warlords were not only pillaging the countryside but were also establishing tollbooths along every road leading to the marketplace in Kabul. Charging exorbitant tolls for safe passage was a common practice in Afghanistan, and even the fee of two million afghanis (about US $300) for every truck that wished to enter Kabul could be exacted. But certain things were happening at the tollbooths that could not be overlooked. Toll collectors at one station had dragged young Muslim men and boys from trucks and cars, forced them to undergo mock public marriages, and then sodomized them repeatedly before robbing them of their goods and permitting them to pass through the gate.[4]

Omar recruited fifty Pashtun students from the nearby *madrassahs* and set out on horseback, like Wyatt Earp with a posse from Dodge City, to set matters straight. They destroyed the tollbooth and killed the sodomites.

The mullah heard reports of other outrages. Two warlords had allowed their followers to repeatedly rape two teenage girls they had kidnapped from Kandahar Province. Omar rallied the students, raided the camp of the warlords, and hanged the offenders from the barrel of a tank.[5]

The defining act for the chivalrous Omar and his faithful band of student soldiers occurred in the midst of the Kandahar bazaar, where two warlords, rivals for the affections of a young boy, fired rounds from their tanks at one another, killing scores of innocent bystanders. The mullah and his men descended upon the scene with swords and scimitars in hand to hack off the heads and limbs of the warlords and their bands of brigands and mercenaries.[6]

Omar became a hero of the people. Thousands joined his cause. Many came from religious schools, hotbeds of Deobandi radicalism, in the frontier provinces of Pakistan. The Deobandi movement began in 1866, when a group of Islamic fundamentalists opened Dar-ul-Uloon, a seminary in Deoband, India, to indoctrinate Muslim students in the traditional values of the faith and to foster a system of education that was free from all Western influence. After the creation of Pakistan in 1947, hundreds of satellites of Dar-ul-Uloon began to sprout up throughout the Muslim nation. The Deobandi schools were small and poorly funded. The students, for the most part, came from impoverished families. The period of education was three years and consisted of intense religious indoctrination, including the commitment of the Koran to memory. If the students were illiterate when they entered (as most were), they remained illiterate upon graduation. Those who completed the training became village *mullahs*, qualified to officiate at births, weddings, and burials. They made their living by setting bones (the one practical skill taught at the schools) and selling religious amulets to protect against the evil eye and the nasty legions of *jinn* (invisible spirits), who inhabited the air and infected women and children with disease.[7] By 1996, 2,512 Deobandi *madrassahs* with a student population of two hundred or more had mushroomed along Pakistan's border with Afghanistan.[8] Since most of Omar's recruits were students (*talibs*) from these schools, his movement became known as the Taliban.

Within a matter of months of its formation, the Taliban had grown into a full-fledged army of thirty thousand. The army received financial support from the royal family in Saudi Arabia, who are Sunni Muslims. They saw in the movement a means to establish a bulwark against the Shiite influence in the region.[9] Additional support came from Inter-Services Intelligence (ISI), Pakistan's secret service agency, which viewed the emergence of the Taliban as a way to manipulate events within the neighboring country.[10]

The United States could not resist yet another opportunity for clandestine military involvement, thinking that Mullah Omar and his gang could serve as a stabilizing force against the extremists in Iran. There was another reason for US interest. American firms wanted to lay a massive pipeline to pump natural gas from Turkmenistan through Afghanistan to Pakistan, a project that could not take place without a dominant central regime in place to suppress the warlords.[11] By late 1994, the CIA through its ISI contacts provided the Taliban with satellite information showing the secret location of abandoned Soviet arsenals that housed tons of arms and ammunition along with fleets of tanks and trucks.[12] Without this information and the discovery of these arsenals, Mullah Omar could not have mounted his attempt to conquer Afghanistan.

By January 1995 the Taliban gained control of the provinces of Uruzgan and Zabol, along with Helmand, Afghanistan's major poppy-growing center. In this way, they cut off the flow of drug revenue to Gulbuddin Hekmatyar, the country's leading warlord and the founder of the Hezb-e-Islami group. On February 14, Hekmatyar fled from his headquarters in Charasyab, south of Kabul, to Iran, leaving behind his stockpile of rocket-propelled grenades (RPGs), machine guns, and Stinger surface-to-air missiles. The Taliban now governed much of Afghanistan, including the sole non-Iranian route between the Indian Ocean and central Asia—a route that would become the financial lifeline of al Qaeda.[13]

Eight months of bloody battle later, the Taliban conquered the capital city of Kabul, driving from power Afghan president Burhanuddin Rabbani and his military commander, Ahmad Shah Massoud. As soon as victory was achieved, Mullah Omar ordered his soldiers to hunt down Muhammad Najibullah, the Afghan communist who had ruled the country from 1986 to 1992. Najibullah took refuge in a UN compound and sent the following communiqué to the Clinton administration: "We have a common task— Afghanistan, the USA and the civilized world—to launch a joint struggle against Islamic fundamentalism. If such fundamentalism comes to Afghanistan, war will continue for many years. Afghanistan will be turned into a center for terror."[14] The plea was ignored. Ignoring diplomatic courtesies, Omar and his soldiers dragged Najibullah from the compound, beat and castrated him, and placed him before a firing squad. His mutilated body, along with that of his executed brother, was left hanging from a light pole for public display in downtown Kabul.[15]

General Massoud and his soldiers withdrew to the Pashir Valley, where they regrouped to form the Northern Alliance, the only pocket of resistance to Mullah Omar's complete control of the country.[16] Massoud's position was fortified when he gained the support of the ousted warlord Hekmatyar, who became his prime minister. The struggle between the Taliban and the Northern Alliance was to continue unabated even after the assassination of Massoud by two al Qaeda agents posing as TV journalists on September 9, 2001, through the launching of Operation Enduring Freedom (America's code name for the "war on terror"), and beyond.[17]

Under Taliban rule, Afghanistan became renamed as the Islamic Emirate of Afghanistan, and *shariah* (Islamic law) in its strictest form was imposed upon the people. All forms of Western technology were outlawed. The *talibs* raised massive bonfires of books, videocassettes, VCRs, and television sets in the marketplaces of towns and villages. Records, cassettes, and compact disks of American music were confiscated and crushed underfoot. Newsstands and bookstores were torched. Movie theaters were shut down. Every form of earthly pleasure—even the flying of kites—became prohibited. Women, dressed in full burqas with mesh covering their eyes, were reduced to the role of breeders and slaves. Men, required to grow full-length beards, were forced at gunpoint to appear at mosques for prayer five times a day.[18] Within this Islamic utopia, sexual intercourse outside of wedlock was punishable by one hundred lashes (for men) and stoning to death (for women), and those who engaged in homosexual acts were crushed by the toppling of brick walls upon their naked bodies. The hands and feet of thieves were amputated in a stadium at Kabul every Friday at 3:30 PM to the cheers of spectators.[19] Public punishment became the primary form of public amusement. Bin Laden had found his Shangri-La.

From the time of his arrival, bin Laden was treated as an honored guest by the Taliban. According to Islamic journalist Abu Abdul Aziz al-Afghani, the student soldiers bowed before bin Laden whenever he appeared in public. One Taliban commander greeted the emir by saying: "O Sheikh! Our land is not the land of the Afghans, but it is the land of Allah; and our *jihad* was not the *jihad* of the Afghans, but the *jihad* of all Muslims. Your martyrs are in every region of Afghanistan; their graves testify to this. You are between your families and your kinsmen, and we bless the soil that you walk upon."[20]

In the eyes of *talib* leaders, bin Laden appeared to have been sent from heaven. The Taliban were in desperate need. They required experienced fighters for the war against the Northern Alliance. They wanted training in military tactics, including the use of explosives and sophisticated weaponry and the basic techniques of guerrilla warfare. Above all, with no commodity production for export and no industry, they needed money—money to pay the soldiers, to purchase munitions and matériel, and to repair the rotting infrastructures throughout the country. These needs, they believed, could be met by the millionaire leader of the *mujahadeen*, the three hundred al Qaeda operatives who had accompanied him from Sudan, and the vast number of recruits he could summon to aid Mullah Omar in the establishment of the *caliphate*.

On August 23, 1996, in Afghanistan, bin Laden, surrounded by the al Qaeda high command and hundreds of militant Islamists from England, Algeria, Lebanon, Egypt, Iran, Yemen, Pakistan, and Saudi Arabia issued "Declaration of War on the Americans Occupying the Country of the Two Holy Places." Strange to say, the *fatwa*, for them, was a statement of exaltation, of expectation, and almost infectious optimism. It reflected bin Laden's belief that he, by the grace of Allah, had arrived in a land of incredible opportunity where he could rebuild his fortune, forge new alliances for the holy war against the enemies of Islam, and bring about his dream of the American Hiroshima.

Within his thirty-four page *fatwa*, written on an Apple Macintosh,[21] bin Laden called upon all Islam—the righteous Sunni and the unrighteous Shiites—to take part in the great struggle against the United States and Israel:

> The ultimate aim of pleasing Allah, praising His word, instituting His religion and obeying His messenger (ALLAH'S BLESSING AND SALUTATIONS ON HIM) is to fight the enemy, in all aspects and in a complete manner; if the danger to the religion from not fighting is greater than that of fighting, then it is a duty to fight them even if the intention of some of the fighters is not pure, i.e., fighting for the sake of leadership (personal gain) or if they do not observe some of the rules and commandments of Islam. To repel the greater of the two dangers at the expense of the lesser one is an Islamic principle which should be observed. It was the tradition of the people of the Sunnah to join and fight with the righteous and

non righteous men. Allah may support this religion by righteous and nonrighteous people as told by the prophet (ALLAH'S BLESSING AND SALUTATIONS ON HIM). If it is not possible to fight except with the help of non righteous military personnel and commanders, then there are two possibilities: either fighting will be ignored and the others, who are the great danger to this life and religion, will take control; or to fight with the help of nonrighteous rulers and therefore repelling the greater of the two dangers and implementing most, though not all, of the Islamic laws. The latter option is the right duty to be carried out in these circumstances and in many other similar situations.[22]

This unified effort was essential since Muslims throughout the world, including bin Laden and his al Qaeda cohorts, believed themselves to be the victims of an alliance of "crusaders" (Osama's favored name for Americans) and "Zionists" (Jews and supporters of Israel)—an alliance that has led to the invasion of Muslim lands (Palestine, Lebanon, Somalia, Iraq), the deaths of millions of innocent men, women, and children as a result of economic sanctions, the wholesale exploitation of the natural resources of Muslim nations, and the establishment of US military bases between the two holy cities. Bin Laden stated further:

We, my group and I, have suffered some of this injustice ourselves. We have been prevented from addressing the Muslims. We have been pursued in Pakistan, Sudan and Afghanistan, which explains my long absence on my part. But by the Grace of Allah, we now have a safe base in the high Hindu Kush Mountains of Khorassan, where, by the Grace of Allah, the largest infidel military force of the world was destroyed. And the myth of the superpower was swept away, while the *mujahadeen* chanted, *"Allahu Akbar"* ("God is great"). Today we work from the same mountains to put an end to the injustice that has been imposed on the Muslim community by the Zionist/Crusader alliance, particularly after the occupation of the blessed land around Jerusalem, the occupation of the very road taken by the Prophet (ALLAH'S BLESSING AND SALUTATIONS ON HIM), and the occupation of the land of the two Holy Places. We ask Allah to grant us victory. He is our Patron and He is the

Most Capable. From here, today we begin the work, talking of correcting what had happened to the Islamic world, in general, and the Land of the two Holy Places, in particular.

Bin Laden concluded the long, rambling *fatwa*, replete with hundreds of quotations from the Koran and Islamic scholars, with the following plea:

My Muslim Brothers of the World:

Your brothers in Palestine and in the land of the two Holy Places are calling upon your help and asking you to take part in fighting against the enemy: the Americans and the Israelis. They are asking you to do whatever you can, with your own means and ability, to expel the enemy, humiliated and defeated out of the sanctities of Islam. Exalted be to Allah, who said in His book: "And if they ask your support, because they are oppressed in their faith, then support them!"(Al-Anfaal 8:72). O you horses [soldiers] of Allah ride on to victory. This is the time of hardship, so be strong. And know that your effort to liberate the sanctities of Islam is the right step toward unifying the word of the *ummah* under the banner of "No God but Allah."

The *fatwa* produced electrifying results. Hundreds of young Muslims traveled to Afghanistan to take part in bin Laden's *jihad*. Training camps for these recruits popped up throughout the countryside. Muslim terrorist groups from countries throughout the world announced their support of al Qaeda. A partial listing of these groups within the new coalition included:

- Afghanistan: Ulema Union of Afghanistan
- Algeria: Armed Islamic Group; Saafi Group for Proselytism and Combat
- Bangladesh: Al-Jihad
- Canada: Groupe Roubaix
- Egypt: al-Gama'a al-Islamiyya; Al-Jihad
- India (Kashmir): Partisans Movement
- Indonesia: Jemaah Islamiyah

- Jordan: Bayt al-Imam; Jihad Organization of Jordan
- Lebanon: Asbat al-Ansar; Hezbollah; Lebanese Partisans League
- Libya: Libyan Islamic Group
- Pakistan: Al-Badar; Harakat ul-Ansar/Mujahadeen; Al-Hadith; Harakat ul-Jihad; Jaish Mohammed; Jamiat Ulema-e-Islam; Jamiat-ul-Ulema-e-Pakistan; Lashkar e-Toiba; Palestinian Authority; Islamic Jihad
- Philippines: Moro Islamic Liberation Front; Abu Sayyaf
- Somalia: Al-Ittihad al-Islami
- Uzbekistan: Islamic Movement of Uzbekistan
- Yemen: Al-Jihad Group[23]

Contributions began to flow into bin Laden's coffers from wealthy Muslim business executives. The money came by courier and through the *hawala* system of interlocking money changers that operates through the use of promissory notes for the exchange of cash or gold.[24] One such business executive was Khalid bin Mahfouz, the CEO of the giant National Commercial Bank in Saudi Arabia. Through this bank, bin Mahfouz funneled more than $100 million to charities that served as fronts for al Qaeda.[25] Similar practices were adopted by the CEOs of other Arab banks, including Saleh Abdullah Kamel of Al-Shamal Bank and Abdullah Suleiman al-Rajhi of the Al-Rajhi Bank. Other large contributors to al Qaeda, included Mohammed Hussein al-Amoudi, the owner of the largest conglomerate in Saudi Arabia; Wael Hamza Julaidan, chairman of the Muslim World League in Pakistan; and Yasin al-Qadi, the president of the Al-Haramain Foundation.[26]

Ironically, millions more came from the Saudi royal family, despite bin Laden's expressed hatred of the regime. Dick Gannon, former deputy director for operations at the US Department of Counterterrorism, offered this simple explanation: "There are certain factions of the Saudi royal family who just don't like America."[27]

In addition to the contributions, radical mosques throughout Saudi Arabia and Middle East began setting aside the *zakat*, the "holy tax" levied upon all believers for worthy causes, for bin Laden.[28] How much al Qaeda received from this source remains anyone's guess. But the amount was surely substantial since wealthy Muslims give extraordinary amounts in order to adhere to the *zakat* and the third pillar of Islam (alms giving). In Saudi Arabia more than $10 billion in *zakat* funds are collected each year.[29]

With these funds, bin Laden was quick to cement his ties with the Taliban. He provided critical help when the Taliban was ready to take over Kabul. The help came in the form of arms, ammunition, fleets of Toyota land cruisers, food, and makeshift hospitals.[30] Bin Laden was now seen on many occasions in the car of Mullah Omar. A hotline was connected between the residences of the two fundamentalist leaders. Osama answered at 925-12-53-06.[31] Eventually, bin Laden built a palace in Kandahar for Mullah Omar, his three wives, and four children. The two became inseparable. Press reports began to surface that Mullah Omar took Osama's daughter as his fourth wife. But these reports have never been substantiated.[32]

By the end of 1996, bin Laden and Mullah Omar embarked on a venture that would secure the future of the newly created Islamic Emirate of Afghanistan, a country without industry or exportable goods. They made a deal with the *bubas*, the Turkish drug lords, and the Albanian Mafia to become the world's leading producer of Number Four heroin, the drug of choice in Europe and the United States. The *jihad* would be fueled by Western decadence, giving truth to the age-old Islamic adage: "It's best to hoist your enemy on his own petard."

CHAPTER THREE

From Albania to the Atom Bomb

It seems clear to me heroin is the Number One financial asset of Osama bin Laden. There is a need to update our view of how terrorism is financed. And the view of Osama bin Laden relying on Wahhabi donations from abroad is outdated. And the view of him as one of the world's largest heroin dealers is the more accurate, up-to-date view.
—Congressman Mark Steven Kirk, 2004

We are making these drugs for Satan—Americans and Jews. If we cannot kill them with guns so we will kill them with drugs.
—Fatwa of Hezbollah

The Afghans are selling 7 to 8 billion dollars of drugs in the West a year. Bin Laden oversees the export of drugs from Afghanistan. His people are involved in growing the crops, processing, and shipping. When Americans buy drugs, they fund the jihad.
—Yossef Bodansky, director of the Congressional Task Force on Terrorism and Unconventional Warfare, 1998

A Binny in Newark, New Jersey, costs twenty dollars. It's a hell of a bargain since the product is 71.4 percent pure. "You can't buy any better heroin in the world than you can in New Jersey," claims Michael Paterchick, special agent in charge of the Newark DEA office.[1]

But to get a Binny for this price, you have to be a regular customer who buys in volume, that is, ten bundles or more. Each bundle contains ten packs or "hits." A hit comes in a small cellophane wrapper that is sealed with an adhesive label. The label bears the Binny brand name and the advertising logo—a murky image of the great emir of the *mujahadeen* with his long beard, prominent nose, dark eyes, and white turban. Unlike the old "nickel" and "dime" bags of hash and grass, these packs have been factory packaged for mass consumption in the manner of old-fashioned penny candy.

Millions of Binnys are sold throughout the United States on college campuses and street corners, in professional office buildings and public restrooms, at rock concerts and movie complexes.

A Binny represents a tenth of a gram of choice Number Four heroin, the finest heroin in the world, which comes straight from the Golden Crescent of Afghanistan, Pakistan, and Iran.

The war on terror has failed to lessen the flow. Although US-led coalition troops continue to occupy Afghanistan, the heroin trade remains the mainstay of the country's economy. Indeed, to the astonishment of all observers, it has soared since the onset of Operation Enduring Freedom (the codename for the 2001 invasion of Afghanistan). In 2004, according to the United Nations, opium poppy cultivation in Afghanistan rose by two-thirds, climbing to 320,000 acres and producing a yield of 4,200 metric tons (a metric ton equals 2,205 pounds).[2] "What we have now is a narco-economy where 40 to 50 percent of the GDP is from illicit drugs," a senior Kabul official told a reporter from the *San Francisco Chronicle*. "The heroin traffickers are naturally interested in supporting terrorism and doing what they can to destabilize the central government because the last thing they want is the establishment of the rule of law. In those terms, it is a matter of national security to the US and Europe."[3]

Few products are more aptly named than "Binnys." Every time a user makes a purchase, he/she is helping to fund the *jihad*. The billions that bin Laden has amassed for the destruction of the United States comes from the sale of this product and the alliance he made with the Taliban, the Turkish *bubas*, and the Albanian or Islamic Mafia.

In 2004 the FBI announced that among ethnic Albanians there is a group of radical Muslims who have replaced La Cosa Nostra (LCN) as the "leading crime outfit in the United States."[4] Few, including national press journalists, took heed of this startling announcement. Most Americans continued to watch *The Sopranos* and to believe that organized crime activities, including prostitution, gun running, labor racketeering, and drug trafficking, remain the domain of established Italian crime families. But this is no longer the case. The Albanian Mafia, owing to their incredible propensity for bloodletting, gained supremacy over the long-entrenched LCN families in every major city along the eastern seaboard. "They are a hardened group, operating with reckless abandon," said Chris Swecker, the newly assistant director of the FBI's Criminal Investigative Division.[5]

The emergence of the Albanian Mafia as the leading force in national organized crime stems, in part, from the FBI's success in busting LCN. As a result of Operation Button-down (the FBI codename for the campaign to crush the Mafia), the number of LCN families shrank from twenty-four to nine. The crackdown resulted in the incarceration of more than one hundred leading members of LCN and six hundred of their associates. By 2001 even the prominent dons of the New York families had been toppled. John Gotti of the Gambino family had died of throat cancer in prison. Vincent Chin Giganti, the "Oddfather" and head of the Genovese family, was sentenced to twelve years in federal prison for murder and labor racketeering. Steven Crea, the leader of the Lucchese family, remained in state prison on charges of enterprise corruption. Alphonse "Allie Boy" Persico, the Colombo family don, was in a federal slammer for gun possession and loan sharking. And Joseph Massino, the Bonnano *capo*, was facing charges of three gangland killings.

The busts left a gap that was filled by the Albanian Mafia. Many had worked as enforcers for the LCN families. Zef Mustafa, for example, had served as the chief "clipper" (hitman) for the Gambinos. Mustafa's capacity for violence was only equaled by his love of liquor. He drank from early morning until late at night, often consuming two to three quarts of vodka a day.[6] While Gotti remained in prison, Mustafa used the resources of the Gambinos to organize a $650 million Internet and phone heist by the use of porno Web sites and 1-800 sex numbers. He was arrested in 2002, pleaded guilty to fraud, and was sentenced to five years in a federal penitentiary.[7]

Abedin "Dino" Kolbiba also served as a hit man for the Gambinos. Dino was so skilled at his craft, including the art of making bodies disappear, that he was contracted out to the other crime families.[8] His present whereabouts remain unknown.

The ruthless indifference of the Albanian Mafia to murder was evidenced by Simon and Victor Deday, two assassins for the Gambinos. The two hit men shot a waiter at the Scores restaurant in New York to express their displeasure at the quality of the service. For good measure, they also shot the bouncer before he could utter a word of protest.[9]

By 1990 radical Muslim émigrés from Albania became the kings of the drug underworld. Skender Fici operated a travel agency in New York that served as a front for a billion-dollar business in narcotics. Fici worked closely with his fellow countrymen, including Ismail Liva. Caught with a stash of more than $125 million in heroin, Fici and Liva issued a contract for a hit on the New York detectives who had arrested them and the federal prosecutors who had placed them in prison.[10]

Daut Kadriovski, yet another member of the Albanian Mafia, escaped from a German prison in 1993 to establish a heroin business in the Bronx. By 1998 Daut's business became so lucrative that he set up branches in Trenton, Philadelphia, Washington, DC, Richmond, Detroit, Chicago, and other cities throughout the country.[11]

In addition to the Italians, other mobsters confess to being afraid of the Albanian Mafia. Speaking anonymously to a reporter from Philadelphia's *City Paper*, a member of the "Kielbasa Posse," an ethnic Polish mob, said that the Poles were willing to conduct business with "just about anybody," including Dominicans, blacks, Hispanics, Asians, and Russians but refused to go anywhere near the Albanian Mafia. "The Albanians are too violent and too unpredictable," the Pole said.[12]

The Albanian Mafia has also gained ascendancy over all other criminal organizations in Europe, where it has become the chief perpetrator of drug smuggling, counterfeiting, passport theft, forgery, trafficking in human body parts, sex slavery, abduction, and murder. Cataldo Motto, Italy's top prosecutor, maintains that members of the Albanian Mafia pose a threat not only to the people of his country but to all of Western civilization.[13]

Dusan Janjic, coordinator of the Forum for Ethnic Relations in Serbia, has offered the following three reasons for the success of the Albanian Mafia: "Firstly, they speak a language that few understand. Secondly, its internal organization is based on family ties, breeding solidarity and safety. Thirdly, there is the code of silence and it is perfectly normal for somebody to die if he violates the code."[14]

But Janjic has overlooked the fourth and most important reason. Many members of the Albania Mafia are devout Muslims who believe that their crimes are serving a religious purpose. This fact became evident when the Italian police began to monitor the telephone conversations of Agim Gashi, Milan's leading drug lord. In one conversation, Gashi told his Turkish suppliers to continue sending shipments of heroin during the holy month of Ramadan. This violation of religious rules is necessary, he said, for the sake of the ultimate goal, namely, "to submerge Christian infidels in drugs."[15] Similarly, in another conversation, Gashi lauded the success of his billion-dollar business by saying: "We have discovered that drugs are not only a source of wealth but a tool to weaken Christendom."[16]

The movement of the *mujahadeen* into the Balkans began in 1992 when the Party of Democratic Action, the ruling Islamic party of

Bosnia, came to power and sent out a call for recruits. Hundreds responded to the call. The recruits were given journalists' credentials to avoid detection by UN officials. Some immediately married Bosnian Muslim women and joined the ranks of the Bosnian army. By 1995 the number of recruits from Islamic countries throughout the world exceeded six thousand.[17]

In 1994 bin Laden, while still living in Sudan, obtained a Bosnian passport from the Bosnian embassy in Vienna and visited an al Qaeda base in Zenica, Albania, that had been organized by al-Zawahiri. The base was an isolated farmhouse that had been converted into a "research center" for advanced weaponry. Much of the research centered on the making of human bombs—individuals who could carry and spread an incredibly virulent form of bubonic plague that remained resistant to treatment with chloramphenicol or one of the tetracyclines.[18] During his visit, bin Laden supervised the creation of an al Qaeda training and command center in Tropje on the property of the Albanian premier Sali Berisha and the establishment of cells in Croatia; Tetovo, Macedonia; and Sofia, Bulgaria, where he met with the Turkish *bubas*.[19]

With the blessing of the *bubas*, bin Laden also set up a laboratory in Hilat Koko within the Turkish-held region of northern Cyprus for the testing of nuclear material, including the enriched uranium that had been secured for him by Jamal al-Fadl.

In 1996, after solidifying his position with the Taliban, bin Laden made another trip to Albania. This time he met with the leaders of his Balkan cells in order to usurp the control of the global narcotics trade from the Sicilian Mafia, who had become partners with the *bubas* in the heroin trade.[20]

For nearly forty years, the *bubas* controlled the flow from Afghanistan through northern Iran and Turkey to Sofia, the capital city of Bulgaria, while the Sicilians transported the product to Greece and Yugoslavia for shipping to western Europe and elsewhere. It was a highly lucrative arrangement. For decades, on any given day, one could find agents of such illustrious Sicilian dons as Luciano Liggio, Salvatore Riina, Bernardo Provenzano, Nitto Santapaola, and the Farrar brothers negotiating terms of the latest shipments from Afghanistan in the opulent lobby of the Hotel Vitosha in Sophia.[21]

But these heady days had come to an end. The Sicilians began to encounter difficulty moving the shipments from Bulgaria because of the prolonged conflict in what once had been Yugoslavia. To make matters worse, the key suppliers of the product—the powerful Afghan warlords, including Gulbuddin Hekmatyar—had been driven from power by Mullah Omar and the Taliban.

In the midst of these developments, the Albanians began to take over the drug trade by carving out a new smuggling route into western Europe that would bypass the peninsula's war zones. Known as the Balkan route, it ran from Bulgaria through Kosovo and into Albania, Greece, and Yugoslavia. From there, the route continued through Europe and to the United States.[22] In 1990 the Hungarian police confiscated 14 pounds of heroin from Albanian smugglers. By 1994 the amount rose to 1,304 pounds.[23]

The Sicilian Mafia, like its counterpart in the United States, became, by and large, a thing of the past, a power that had been spent. Proof of this was evidenced by the ability of the Albanians to infiltrate and gain control of various branches of the Sicilian organization, including Sacra Corona Unita ("Holy United Crown") in Apulia, just across the Otranto Canal from Albania.[24] And they took over the prostitution racket in Italy before the once-great dons could utter a word of protest, let alone squeeze a trigger.[25]

During his 1996 visit, bin Laden forged a working relationship between the Albanians, the Turks, and the Taliban. The arrangement was suitable to the *bubas,* who were pleased to work with fellow Muslims rather than Christian infidels.

Bin Laden realized the precariousness of reliance upon one supply route from Afghanistan to the West. The flow of product along this route remained contingent upon the cooperation of Iran and a degree of stability in Kosovo. And so, he made arrangements with the Chechen brothers, Shamil and Shirvani Basayev, to reestablish the so-called Abkhaz drug route.[26] This Abkhaz route led from Afghanistan through Khorog in Tajikistan and Osh in Kyrgyzstan along a treacherous mountain pass to the Vedensky Rayon in Chechnya. From this location, Abkhaz trucks and MI-6 helicopters would transport the heroin to the port city of Sokhumi in the breakaway Georgian province of Abkhaz, where it would be loaded on Turkish ships and shipped under the watchful eye of the *bubas* to the port of Famagusta in northern Cyprus. Here it also would be broken down into small packets by the Albanian Mafia for distribution throughout the world.[27]

As soon as bin Laden returned to Afghanistan, heroin began to flow into Sofia and Famagusta at the rate of six metric tons a month.[28] The international market rapidly became saturated. Business boomed, while prices dropped. By the end of 1996, prisons throughout Europe became filled with drug dealers from the Balkans. In Germany alone, more than eight hundred Albanians were incarcerated on possession of an average of 124 grams of choice Number Four.[29] Tony White of the UN Drug Control Program observed: "They [the Albanian mob] are more likely to use violence than any other group. They have confronted the established order throughout Europe and pushed out the Lebanese, Pakistani, and Italian cartels."[30]

Prior to bin Laden's exile from Sudan, the heroin that came out of the Golden Crescent was low-grade Number Three. It was a product that was good only for smoking or snorting. The production process of heroin began when the petals from the poppies fell, exposing egg-shaped seed pods. The pods were sliced vertically to extract the opaque, milky sap that represented opium in its crudest form. As it darkened and became thicker, the extracted sap became compacted into bricks of opium gum for transport to refineries. At the refineries of the Hekmatyar and those of other warlords, the opium gum was mixed with lime in huge vats of boiling water. A precipitate of waste sank to the bottom while the white morphine rose to the top. The morphine was skimmed from the vats, reheated with ammonia, filtered, and boiled again until it became reduced to a brown paste. The paste was poured into mounds and dried in the sun. The finished product was now ready for transport to the *bubas*.

But the market now called for Number Four that could be mainlined into the veins for the ultimate high or skin-popped by yuppies for an experience that was "better than sex." Skilled chemists working in sophisticated laboratories were required for the production of this product. The opium paste had to be placed back into vats, reheated, and processed with acetic anhydrite for six hours in glass containers to form diacetylmorphine. The solution was drained and sodium carbonate added to make the heroin solidify and sink. The heroin became further refined by filtering it from the sodium carbonate solution through activated charcoal and purifying it with alcohol. The product was now reheated to evaporate the alcohol. Next came the tricky part as the heroin was mixed with ether or hydrochloric acid, a process that could produce a violent explosion if handled by an improperly trained technician. The end result was Number Four heroin—a fluffy white powder that would be shipped in twenty- to one-hundred kilo packages.

To meet the demand for this product, bin Laden established sophisticated laboratories near Kabul and recruited chemists from Pakistan, China, and the former Soviet Union.[31]

Within weeks, the facilities were up and running. Afghan farmers were now ordered by Mullah Omar to produce "all the opium they could grow."[32] The results were astounding. By 1997 the poppy harvest soared to 3,276 tons of raw opium and revenues began to pour into the Taliban treasury at a rate estimated from $5 billion to $16 billion a year.[33]

A narcotics nightmare had been unleashed. By 1998 the National Household Survey discovered 149,000 new heroin users throughout the United States who required treatment for their addiction. Eighty percent of the new addicts were under the age of twenty-six. The average addict was spending from $150 to $200 a day to maintain his habit.[34] While the heroin trade increased to new heights in the United States, it began to soar beyond belief in Europe. From 1996 to 2001, Europeans consumed more than fifteen tons of heroin a year—twice the amount sold in America—and almost all of it (over 90 percent) came from Afghanistan.[35]

Although teachings of Islam are opposed to intoxicants,

including alcohol and narcotics, opium now became traded at bazaars and marketplaces throughout the Islamic Emirate of Afghanistan. One of the best places to obtain heroin at a rock-bottom price turned out to be the town of Sangin, a three-hour drive from Kandahar. More than 250 shops that lined the main street sold wet opium in plastic bags and dry opium in large cakes. One trader boasted to UN officials of selling sixty-two thousand pounds a year and amassing a gross income of $132,000.[36]

A double standard came to exist within the Taliban government. While the religious militia, with leather whips in hand, tirelessly patrolled the streets in pursuit of violators of the *shariah*, the stalls of opium dealers remained open on every street corner. While executioners beheaded apostates, infidels, rapists, and murderers in the stadiums, Afghan drug merchants were permitted to sell their wares and become rich. The new drug lords drove their all-terrain vehicles, imported at great expense, through the miserable streets of Kandahar and Kabul that had become open cesspools, replete with rancid garbage and raw sewage.[37]

The stated purpose of the trafficking, according to the *muja-hadeen*, was the poisoning of the *kafirs* in the Western world. But there proved to be a blowback for the Islamic nations. By 1999 there were more than two million heroin addicts in Pakistan and more than one million in Iran.[38]

On July 27, 2000, Mullah Omar announced that his government had decided to ban the growing of opium poppies within the Islamic Emirate of Afghanistan. The United Nations applauded this announcement, claiming that it represented a major step in the effort to curb the sale and distribution of heroin. After the enactment of the prohibition, a handful of farmers were arrested and their fields were destroyed. But the prohibition was merely a ploy to placate world opinion and to gain UN recognition of the Islamic Emirate of Afghanistan. NATO satellite surveillance showed that more surface areas of farming throughout Afghanistan were devoted to growing opium poppies in 2000 than any other year.[39]

CHAPTER FOUR
The Three Wars

The war on terrorism skipped the KLA. Kosovo has become exclusively an Albanian Province. . . . The Balkans, since the end of the bombing, have been in constant turmoil caused by the KLA terrorist activities.
　　—James Bissett, former Canadian ambassador to Yugoslavia,
November 13, 2001

The towers are an economic power and not a children's school. Those that were there are men that supported the biggest economic power in the world. They [the American people] have to review their books. We will do as they do. If they kill our women and our innocent people, we will kill their women and their innocent people. Not all terrorism is cursed; some is blessed. America and Israel exercise their condemned terrorism. We practice the good terrorism that stops them from killing our children in Palestine and elsewhere.
　　—Osama bin Laden, November 2002

The enemy in this war is not "terrorism" but militant Islam.
　　—Eliot Cohen, the *Wall Street Journal*, November 20, 2001

The billions amassed from the drug trade were used by al Qaeda to fund three wars. The first was the Taliban's struggle against the Northern Alliance, which proved to be costlier than expected since Ahmad Massoud, the "Lion of the Panjshir," was a gifted strategist. Every year the Taliban forces would set out across the Shomali plains to do battle with Massoud's forces but never managed to achieve a decisive victory. Finally, as noted earlier, on September 9, 2001, bin Laden dispatched three assassins to the headquarters of the Northern Alliance in the northern village of Khvajeh Boha. The assassins gained entry to Massoud by posing as journalists and set off a bomb, killing the celebrated general and four of his guards.[1]

The second war funded by heroin was the conflict in Kosovo. In 1997, bin Laden paid a third visit to Albania to aid in the creation of the Kosovo Liberation Army (KLA). The purpose of the KLA was threefold: (1) to reverse the efforts and to combat the war crimes of the Christian Serbs under Slobodan Milosevic in purging Kosovo of its Albanian population; (2) to undermine the attempt of Ibrahim Rugova and the League for Democratic Kosovo to obtain a peaceful resolution to the struggle between the Serbs and ethnic Albanians; and (3) to unite the Albanian populations of Kosovo, Macedonia, and Albania into Greater Albania.

To accomplish these objectives, bin Laden provided between $500 and $700 million and the expertise of five hundred seasoned Arab Afghan troops to train KLA recruits at the al Qaeda central Balkan headquarters in Tropje, Albania, and a training camp in neighboring Macedonia. The Macedonian camp, located in the mountains of Spopaska Crna Gora near Skopje, was placed under the command of Zaiman al-Zawahiri, brother of Ayman al-Zawahiri.[2]

At this point in the twisted history of Kosovo, the CIA and the Clinton administration began to view the KLA as an army of "freedom fighters" and offered aid in the form of military training and field advice.[3] The United States, unbeknown to the American people, was now in league with a group that contained enemies who were intent upon its destruction. There were generally not the innocent people who had been targeted and attacked by the Serbs.

By 1998 the KLA became a formidable army of thirty thousand with sophisticated weaponry including shoulder-fired antitank rocket launchers, mortars, recoilless rifles, and antiaircraft machine guns. From Tropje, the KLA began to conduct hit-and-run attacks on Serbian special-forces police units. Milosevic and the Serbian officials responded by burning homes and killing dozens of ethnic Albanians in the Drenica region of Kosovo. A full-scale conflict erupted, culminating in the infamous "Racak Massacre" of January 15, 1999, when the bodies of forty-five Albanians were discovered in a gully within the village of Racak.[4]

Confronted with this crime, Milosevic insisted that the bodies had been placed in the gully by the KLA in order to implicate the Serbs and to justify Western intervention. *Le Figaro* and *Le Monde* suggested that his claim was supported by the unnatural position of the bodies, the absence of cartridge shells, and the inability of the Racak villagers to identify the victims.[5]

In the wake of outcries of genocide and ethnic cleansing, the United States became involved. Both sides were asked to sit down and negotiate a settlement. Members of the Clinton administration produced the paperwork and asked everybody to sign it. Among other things, the agreement called for a referendum on independence for Kosovo. Milosevic refused. The United States, with its NATO allies, responded by reducing Kosovo to a heap of rubble. Between March 24 and June 10, 1999, a total of 37,465 missions were flown, destroying 400 Serbian artillery weapons, 270 armored personnel carriers, 150 tanks, and 100 planes; killing 10,000 Serbian soldiers and a convoy of refugees; and causing 1.4 million Kosovars to flee for their lives—the greatest mass migration since World War II. The cost of the weeks of bombardment exceeded $4 billion. Milosevic and the Serbs decided that they had enough. The accord of June 21, 1999, ended the bombing, eliminated the presence of Milosevic's regime in Kosovo, and authorized a NATO force of 1,700 police officers to retain order until democratic elections could be held.

But the situation in Kosovo was far beyond the capability of the meager NATO police force. Hashim Thaci, a leader of the KLA, now proclaimed himself prime minister of the Kosovo government. The ethnic Albanians, who had fled for safety from the Serbs, now returned with a vengeance. It became an eye for an eye, and neither the Clinton administration nor the United Nations uttered cries of alarm. More than two hundred Christian churches and monasteries were destroyed. Many, including the Devic Monastery, the Cathedral of St. George, the Church of St. Nicholus, and the Monastery of the Holy Archangels, had been built in the thirteenth and fourteenth centuries.[6] Reports surfaced of mass executions of Serbian farmers, the murder of scores of priests, and "granny killings"—the drowning in bathtubs of elderly women.[7] Of the forty thousand Serbs who lived in Kosovo before the war, only four hundred were left within a month after Kosovo became a NATO protectorate.[8] The Serbs who remained were sequestered in three gloomy apartment buildings on the north side of town, where the international police stood guard night and day.[9] By September 1999, over two hundred thousand Serbs and thirty thousand gypsies had been exiled from the country.[10]

Despite the best intentions of the Clinton administration, bin Laden and the more radical members of the KLA had won a decisive victory. Hundreds of Wahhabi mosques and schools, thanks to contributions from wealthy Saudis, now appeared in every town and village throughout the country. Kosovo, with a Muslim population of 1.8 million, now stood as an Islamic bulwark in the midst of the Balkans.[11] Although many Muslims and their families simply wanted to eke out a postwar existence and get on with their lives, the result of the war had done something more for bin Laden. The backdoor to Europe had been pried open for the drug trade and the movement of weapons of mass destruction.

The third war funded by the drug trade was the great *jihad* against the United States of America. In February 1998, on behalf of the

International Islamic Front, bin Laden issued his second *fatwa* against the United States with the following declaration: "The ruling to kill the Americans and their allies—civilians and military—is an individual duty for every Muslim who can do it in any country in which it is possible to do it, in order to liberate the Al-Aqsa mosque and the holy mosque [Mecca] from their grip, and in order for their armies to move out of all the lands of Islam, defeated and unable to threaten any Muslim. This is in accordance with the words of Almighty Allah, 'Fight the pagans all together as they fight you all together,' and 'Fight them until there is no more tumult or oppression, and then will prevail justice and faith in Allah.'"

In the declaration, he presented three reasons why Muslims throughout the world must unite with al Qaeda in this holy war:

First, for over seven years the United States has been occupying the lands of Islam in the holiest of places, the Arabian Peninsula, plundering its riches, dictating to its rulers, humiliating its people, terrorizing its neighbors, and turning its bases in the Peninsula into a spearhead through which to fight the neighboring Muslim peoples. If some people have in the past argued about the fact of the occupation, all the people of the Peninsula have now acknowledged it. The best proof of this is the Americans' continuing aggression against the Iraqi people using the Peninsula as a staging post, even though all its rulers are against their territories being used to that end, but they are helpless.

Second, despite the great devastation inflicted on the Iraqi people by the crusader-Zionist alliance, and despite the huge number of those killed, which has exceeded 1 million, the Americans are once again trying to repeat the horrific massacres, as though they were not content with the protracted blockade imposed after the ferocious war or the fragmentation and devastation. So now they come to annihilate what is left of this people and to humiliate their Muslim neighbors.

Third, if the Americans' goals behind these wars are religious and economic, their aim is also to serve the Jews' petty state and divert attention from its occupation of Jerusalem and murder of Muslims there. The best proof of this is their eagerness to destroy Iraq, the strongest neighboring Arab state, and their endeavor to fragment all the states of the region such as Iraq, Saudi Arabia, Egypt, and Sudan into paper statelets and through their disunion and weakness to guarantee Israel's survival and the continuation of the brutal crusade occupation of the Peninsula.[12]

On August 7, 1998, bin Laden gave vivid proof of the earnestness of his declaration of war by ordering the bombings of the US embassies in Kenya and Tanzania, killing 234 people, twelve of them Americans, and wounding five thousand others. His plan to bomb a third US embassy—this one in Kampala, Uganda—was prevented by a tip from a CIA informant.[13]

Thirteen days later, the US forces retaliated by firing a host of cruise missiles at al Qaeda residential and military complexes in Khost, Afghanistan, and by the bombing of the al-Shifa Pharmaceutical Plant near Khartoum, believing it to be bin Laden's laboratory for the production of the deadly nerve gas VX.[14] But bin Laden, al-Zawahiri, and other members of the al Qaeda high command were not in the camps at Khost. They were safe and secure in a Pakistani *madrassah*. Although the missiles did strike the camps, the only casualties were local farmers and some low-level militants.[15] The camps themselves had been constructed of stone, wood, and mud, making it easy for the al Qaeda operatives to rebuild them in a matter of days. And the Sudan factory was not involved in the production of VX or any other chemical weapon. The plant simply produced common pharmaceuticals, including ibuprofen.[16]

The US response to the bombings of the embassies was more than a resounding dud—several of the bombs even failed to detonate upon impact. It proved to be a rallying call to Muslims throughout the Middle East. The Afghan camps contained five mosques to provide religious services not only to the *mujahadeen* but also to the local villagers. Four mosques were destroyed, leaving the area littered with the burned pages of hundreds of Korans. Pictures of the bombed mosques and the burned pages were circulated throughout the Muslim world. "America has invited death upon itself," Mauvi Fazlur Rehman Khalil, head of Harakat ul-Ansar, told the press: "If we don't get justice from the world court, we know how to get our own justice."[17]

The missile attacks, in the final analysis, only served to enhance bin Laden's image as a righteous warrior. Rahimullah Yusufzai, the Pakistani journalist who interviewed bin Laden for ABC News, noted: "In an Islamic world desperately short of genuine heroes, Osama bin Laden has emerged as a new cult figure."[18] The Sudanese spiritual leader Hassan Abdallah al-Turabi told the *Christian Science Monitor*: "Bin Laden lives in a very remote place, but now—ho, ho—you [Americans] raised him as the hero, the symbol of all anti-West forces in the world. All the Arab and Muslim young people, believe me, look to him as an example." Al-Turabi concluded his remarks by saying that the attacks would serve to create "100,000 bin Ladens."[19]

Next bin Laden focused his attention on preparations for a "night of power" that would occur on January 3, 2000, the holiest day of Ramadan, with the bombing of a US warship, USS *The Sullivans*. But al Qaeda's small boat, loaded with boxes of heavy explosives, sank as soon as it was launched, forcing the emir to adopt alternative plans. These plans came to fruition on October 10, 2000, when USS *Cole* cruised into Aden in Yemen for a "gas and go" stop. As soon as the *Cole* was anchored, a skiff packed with five hundred pounds of C-4 plastic explosives sped across the harbor toward the warship. The US soldiers waved at the terrorists who were manning the boat. As soon as the skiff reached the port side of the hull, the terrorists pressed the detonator, blowing a gaping hole in the side of the *Cole*. Seventeen American soldiers were killed. The only disappointment for bin Laden was that the al Qaeda operative who was ordered to videotape the attack for propaganda purposes fell asleep at his observation point on Steamer Point.[20]

In the wake of the attack, the Clinton administration waved its sword at bin Laden and pledged revenge. But no retaliatory measures were adopted. This lack of action was accepted as another sign of American "weakness," emboldening bin Laden to launch his next and most audacious attack.[21] In his mind, the skiff symbolized al Qaeda and the *Cole* represented the United States of America.

September 11, 2001, represented an attack of amazing precision. At 8:46 AM, American Airlines Flight 11 smashed into the north tower of the World Trade Center between the ninety-fourth and ninety-eighth floors at the speed of 470 miles per hour. The Boeing 767-200 had departed from Boston's Logan Airport with ninety-two passengers and ten thousand gallons of fuel.

At 9:02 AM, United Airlines Flight 175 hit the south tower between the seventy-eighth and eighty-fourth floors at the speed of 586 miles per hour. The aircraft, also a Boeing 767-200 from Boston, carried seventy-five passengers and ten thousand gallons of fuel.

At 9:40 AM, American Airlines 77 struck the west side of the Pentagon. The Boeing 757 had departed from nearby Dulles Airport with sixty-five passengers. The plane blasted an opening five stories high and two hundred feet wide in the military complex.

There was only one disappointment for the planners. At 10:10 AM, United Airlines Flight 93 failed to strike the White House, which had been its target. Instead, it crashed in a field in Shanksville, Pennsylvania, eighty miles southeast of Pittsburgh, with forty-five passengers aboard. The 757 had been scheduled to depart from Newark Airport at 8:00 AM, but a forty-minute delay put the flight behind schedule and placed the mission in jeopardy. Through cell phones, passengers learned of the attacks on the World Trade Center and the Pentagon and knew what awaited them when the al Qaeda operatives attempted to seize control of the plane. A struggle ensued between the passengers and the hijackers, resulting in the unfortunate crash. The mishap had occurred not by any fault in planning or weakness of resolve but simply because of a mere quirk of fate.[22]

The Twin Towers had stood as an abomination before the eyes of bin Laden and the *mujahadeen*. They loomed as huge monuments to mammon, architectural crystallizations of Western greed and imperialism, and modern-day Towers of Babel that rose as effronteries before the throne of Allah. Their fall was greeted by shouts of joy from millions of Muslims in Indonesia, Malaysia, Bangladesh,

India, Egypt, Sri Lanka, Oman, Yemen, Sudan, Bosnia, Afghanistan, Pakistan, and the United Kingdom.[23]

On September 20, 2001, President George W. Bush addressed the nation concerning plans for the upcoming war on terror as follows:

> This war will not be like the war against Iraq a decade ago, with a decisive liberation of territory and a swift conclusion. It will not look like the air war above Kosovo two years ago, where no ground troops were used and not a single American was lost in combat.
>
> Our response involves far more than instant retaliation and isolated strikes. Americans should not expect one battle, but a lengthy campaign, unlike any other we have ever seen. It may include dramatic strikes, visible on TV, and covert operations, secret even in success.
>
> We will starve terrorists of funding, turn them one against another, and drive them from place to place, until there is no refuge or no rest. And we will pursue nations that provide aid or safe haven to terrorism. Every nation, in every region, now has a decision to make. Either you are with us, or you are with the terrorists.
>
> From this day forward, any nation that continues to harbor or support terrorism will be regarded by the United States as a hostile regime.
>
> Our nation has been put on notice: We are not immune from attack. We will take defensive measures against terrorism to protect Americans.[24]

Operation Enduring Freedom, the codename for the US-led invasion of Afghanistan, began on October 7, 2001, with a salvo of fifty missiles from three US cruisers and a US destroyer in the Arabian Sea. One missile struck the residence of Mullah Omar in Kandahar, killing his stepfather and his ten-year-old son. Another hit the small village of Kouram, killing one hundred civilians. A third destroyed a mosque and a residential village near Jalalabad.[25] By November 9, the coalition forces gained control of the strategic town of Mazar-i-Sharif, putting an end to the rule of the Taliban. Two days later, Taligan in the south and Herat in the north fell with hardly a shot being fired. On November 12, the coalition forces entered Kabul to cheers from a carefully choreographed crowd of onlookers.[26]

The spirits of the US military officials and the Bush administration soared. The invasion was expected to be a dangerous venture fraught with guerrilla attacks and weapons of mass destruction. But it had turned out to be a cakewalk.

The only alarming development came from a discovery made within a house in the upscale neighborhood of Wazir Akbar Khan in Kabul, where the al Qaeda top command resided. Several seasoned war correspondents had found documents containing instructions on how to prepare chemical weapons from common household items, including the production of deadly ricin from castor oil seed. They also came upon instructions, replete with illustrations, for the construction of dirty nuclear bombs and a charred twenty-five page document. Written in Arabic, German, Urdu, and English, this document called "Superbomb" showed how TNT could be used to compress plutonium into a critical mass, sparking a chain reaction that would lead to a thermonuclear explosion.[27] David Albright, a former nuclear weapons inspector, examined the documents and concluded that "the author understood shortcuts to making crude nuclear explosives."[28]

The coalition forces pressed on to Jalalabad, where they continued to encounter little resistance. During the first six weeks of combat, the only US fatalities occurred on November 10, when the Taliban managed to shoot down a helicopter, killing two airmen. The incident occurred in Pakistan near the Dalbandin air base, fifty miles from the Afghan border.[29]

Believing that bin Laden was holed up in an elaborate mountain fortress in Tora Bora, the highest peak of the Spin Ghar or "White Mountains," with two thousand operatives, the United States began a massive air strike with "bunker blasters" and "daisy cutters." It unleashed these 6,800-kilogram bombs—the largest conventional weapons in the US arsenal—with the intent of pounding the mountain lair of al Qaeda to smithereens.[30] When the bombing ended and the last cave was breached, the military expected to find hundreds of bodies and to take hundreds of prisoners. But few bodies were found among the rubble and only nineteen emaciated and toothless captives could be rounded up for a parade before the international press in Kandahar.[31]

As they combed the tunnels near a former al Qaeda base in Kandahar, US troops came upon another discovery that told of bin Laden's plans for the future: low-grade uranium-238 in a lead-lined canister. Although not weapons grade and unsuitable for use in the construction of a fission bomb, the uranium could be combined with conventional explosives to produce a "dirty nuke."[32] Such a device, if detonated in a major American city, could kill thousands of people, contaminate tens of thousands more, cause a mass evacuation, and send the US economy into a tailspin from which it might not fully recover. The fact that the retreating al Qaeda fighters were willing to leave behind such valuable nuclear material—worth millions on the black market—gave rise to increased speculation that they must have taken their "crown jewels"—a small arsenal of tactical nuclear weapons—with them.[33]

This conviction was fortified by a confidential report from British intelligence that told of two special agents who, in 2001, had infiltrated an al Qaeda training camp in Afghanistan by posing as recruits from a London mosque. After weeks of intensive training in guerrilla warfare coupled with religious indoctrination, the two agents swore the *bayat* (oath of allegiance) and were sent to Herat in western Afghanistan for special operations. In Herat, they visited an al Qaeda laboratory where scientists and technicians were busy putting the finishing touches on a sophisticated nuclear weapon that they had developed from radioactive isotopes. This weapon, let alone others that had been produced by the terrorist organization, has never been discovered.[34]

The spring offensive in Afghanistan was launched on March 2, 2002. Called "Operation Anaconda," it was based on false information that thousands of al Qaeda and Taliban fighters were regrouping in the mountainous region of Shah-i-Kot to strike US military bases and to topple the interim Afghan government of Hamid Karzai.[35] To address this threat, megaton bombs were dropped at the rate of 260 a day. On March 8, President Bush said: "These people evidently don't want to give up, and that's okay. If that's their attitude, we'll just have to adjust, and they'll have made a mistake."[36]

On the ground, the coalition forces were reported to have met with heavy resistance. The numbers of "confirmed dead" among the enemy forces, as the *New York Times* pointed out, began to rise and fall like "the fluctuations of a troubled currency: 100, 500, 200, 800, 300."[37]

When the offensive came to an end on March 18, only ten enemy soldiers were taken prisoner, and fewer than twenty bodies were found in the battle zone.[38] The 3,250 bombs had been dropped on largely uninhabited territory.[39]

Operation Anaconda gave way to other offensives—Operation Ptarmigan, Operation Snipe, Operation Condor, and Operation Buzzard—that resulted without the spotting of a single Taliban soldier or al Qaeda operative.

The one success came with the arrest of Khalid Shaikh Mohammed, al Qaeda's military operations chief, in Karachi, Pakistan, on March 2, the launch date of the ill-fated Operation Anaconda. After days of interrogation, the terrorist chief admitted that bin Laden was preparing to create a "nuclear hell storm" in the United States like the 1945 blast in Hiroshima that had killed 140,000 Japanese civilians. Unlike other attacks that could be planned and conducted by lower-level al Qaeda leaders, Khalid Mohammed said, the chain of command for the nuclear operation—"the American Hiroshima"—answered directly to bin Laden, al-Zawahiri, and a mysterious scientist called "Dr. X."[40]

PART TWO

THE CROWN JEWELS

When your Lord revealed to the angels: I am with you, therefore, make firm those who believe. I will cast terror into the hearts of those who disbelieve. Therefore strike off their heads and strike off every fingertip of them.

This is because they acted adversely to Allah and His Apostle; and whoever acts adversely to Allah and His Apostle—then surely Allah is severe in requiting (evil).

This—taste it, and (know) that for the unbelievers is the chastisement of fire.

O you who believe! When you meet those who disbelieve marching for war, then turn not your backs to them.

And whoever shall turn his back to them on that day—unless he turn aside for the sake of fighting or withdraws to a company—then he, indeed, becomes deserving of Allah's wrath, and his abode is hell; and an evil destination shall it be.

—"The Accessions," Koran 8:12–16

CHAPTER FIVE

The Loose Nukes

MR. RUSSERT: *Do you think Osama is still fully in control of al Qaeda?*

MR. SCHEUER: *I think it's wishful thinking to think that he isn't, sir. The one example is the tremendous sophistication and spontaneity of his media machine. There has to be some command and control there. And to imagine that it doesn't—that he's unable to do it is just absolutely incorrect. He's really a remarkable man, a great man in many ways, without the connotation positive or negative. He's changed the course of history. You just have to try to take your fourth-graders' class to the White House visitors' center . . .*

MR. RUSSERT: *When you say "great man," people cringe.*

MR. SCHEUER: *Yes, sir. Absolutely they cringe, but a great man is someone—a great individual is someone who changes the course of history. And certainly in the last five or six years, America has changed dramatically in the way we behave, in the way we travel. Certainly he's bleeding us to death in terms of money. Look at the budget deficit now. Much of that goes against Osama bin Laden.*

MR. RUSSERT: *Do you see him as a very formidable enemy?*

MR. SCHEUER: *Tremendously formidable enemy, sir, an admirable man. If he was on our side, he would be dining at the White House. He would*

be a freedom fighter, a resistance fighter. It's—and again, that's not to praise him, but it is to say that until we take the measure of the man and the power of his words, we're very much going to be on the short end of the stick.
—CIA analyst Michael Scheuer, interviewed by Tim Russert,
Meet the Press, NBC News, November 4, 2004

The beginning of the end took place on September 27, 1991, when President George H. W. Bush announced that the United States would unilaterally withdraw all nuclear weapons from its forces around the world with the proviso that Soviet leader Mikhail Gorbachev would do likewise. Gorbachev readily agreed. The press and the people of both nations were pleased. The agreement represented a grand display of *perestroika*—positive proof that the cold war was over and a new day of mutual trust (if not brotherhood) had dawned. The process of the withdrawal of the nukes was greatly facilitated by the enactment of the Nunn-Lugar Soviet Threat Reduction Act on December 12, 1991. This act provided Russia with millions in funding to transport its nukes to various sites throughout the fourteen newly created republics of what was once the Soviet Union. This, too, seemed like a sensible measure that would make the world a safer place.

Yet the movement of the twenty-two thousand nuclear weapons occurred when everything in Russia was falling apart. With the closing of the communal plants and industries, more than 30 percent of the workforce became unemployed. Inflation soared over 2,000 percent, fueling crime and corruption. An average of eighty-four murders took place a day, many of which were contract killings.[1] A hit could now be arranged with members of eight Mafia families within Moscow for less than $200. Millions of Russians now stood in lines for hours in order to redeem government-issued coupons. There was one line for red beets and cabbages; another for eggs and bread; and yet another for vodka and cigarettes. Butcher shops sold blue chickens that had died of malnutrition and pies made from rancid beef and horse meat. The savvy customers soon learned not to purchase the pies that were surrounded by dead flies. In the wink of an eye, the second most powerful nation on earth had become transformed into a third world country.

Under the new capitalism, male life expectancy fell to fifty-eight years—fifteen to seventeen years less than that of males in sestern Europe and the United States.[2] By 1996 the suicide rate had doubled. Along with poverty, inflation, unemployment, and depression, Russians experienced a sharp drop in the birth rate. From 1991 to 2001, the population declined at a rate of 1.2 million a year.[3]

Nowhere was the misery more apparent than within the military, which by 1996 had shrunk to a feeble force of 1.7 million soldiers. Because of chronic food shortages, many soldiers resorted to begging. More than ten Russian soldiers died each day from noncombat causes, including suicide and malnutrition.[4] An estimated 110,000 lacked proper housing and became sheltered in hovels. No one in the military, not even a high-ranking general, was receiving a regular pay check.[5] Russian army and navy officers began to sell almost every item at their disposal. In 1993 there were 6,430 reports of stolen weapons from army arsenals, ranging from assault rifles to tanks.[6]

It is incredible to assume that the twenty-two thousand nuclear weapons were moved from strategic sites to arsenals throughout Russia without a single loss. When President George H. W. Bush announced his plans for a nuclear withdrawal, the then secretary of defense Dick Cheney said that the recovery of 90 percent of the nukes in Russia would represent "excellent performance."[7] Such an "excellent performance" would mean that 220 weapons would have been lost, stolen, or otherwise unaccounted for. But what person in his right mind could expect such an outcome from the poorly housed, malnourished, disillusioned, and unpaid Russian troops of 1991? The temptation for gain would have been too great for an Orthodox saint to suppress when a kilo of chrominum-50 would sell for $25,000, cesium-137 for $1 million, and lithium for $10 million. Prospective buyers included agents from North Korea, Pakistan, Libya, and a well-financed group of Muslim terrorists called al Qaeda.

In the first three years after the collapse of the Soviet Union, the black market in nuclear weapons and materials began to boom. Germany reported more than seven hundred attempted nuclear sales. This chart contains a partial list of *confirmed* incidents from 1992 to 1995 that involved weapons-grade nuclear material—material that could be used in the construction of a nuclear bomb:

	Source of material	Country where material was seized	Material/ quantity	How material was found
May 1992	Russia (Luch) Scientific Production Assoc.	Russia	1.5 kilograms (90 percent HEU)	Police investigation
March 1993	Russia	Russia	3.05 kilograms (90 percent HEU)	Police investigation
May 1993	Russia	Lithuania	0.1 kilogram (50 percent HEU)	Police investigation
July 1993	Russia	Russia	1.8 kilograms (36 percent HEU)	Police investigation
November 1993	Russia	Russia	4.5 kilograms (20 percent HEU)	Police investigation
May 1994	Un-specified	Germany	0.006 kilogram plutonium-239	Police investigation
June 1994	Russia	Germany	0.0008 kilogram (87.8 percent HEU)	Police investigation
July 1994	Russia	Germany	0.00024 kilogram plutonium	Police investigation
August 1994	Russia	Germany	0.4 kilogram of plutonium	Police investigation
December 1994	Russia	Czech Republic	2.7 kilograms (87.7 percent HEU)	Police investigation
June 1995	Russia	Czech Republic	0.0004 gram (87.7 percent HEU)	Police investigation
June 1995	Russia	Czech Republic	0.017 kilogram (87.7 percent HEU)	Police investigation
June 1995	Russia	Russia	1.7 kilograms (21 percent HEU)	Police investigation

Note: Uranium enriched with 20 percent or higher U-235 is considered weapons-usable material. HEU stands for highly enriched uranium. One kilogram equals 2.2 pounds. One thousand grams equal 1 kilogram, and 1 gram is equal to about 0.04 ounce, or the weight of a paperclip.

Source: GAO Report, May 2002: Nuclear Nonproliferation

These illegal sales were luckily stopped by the police, but that was not true of every case of smuggling. Other incidents have been more alarming. In January 1992 an Egyptian newspaper reported that Iran bought three Soviet nuclear warheads from Kazakhstan for $150 million. This report was later confirmed by Russian intelligence.[8] In November 1993 two nuclear warheads, sufficient to kill millions in New York and Los Angeles, were stolen by two employees from the Zlatoust-36 Instrument Building Plant, a weapons assembly facility, in Chelyabinsk. Fortunately, the weapons were later recovered in a nearby residential garage, and the two employees were placed under arrest.[9] Later that same month, Russian navy Captain Alexei Tikhomirov broke into a nuclear storage facility at the Sevmorput shipyard near Murmansk and robbed three pieces of a reactor core containing 3.4 kilograms of highly enriched uranium. At the time of Tikhomirov's arrest, the chief Russian prosecutor noted that "potatoes were guarded better" than nuclear materials at Murmansk.[10]

The Chechen Mafia, the most powerful criminal organization in Russia, now became the most dominant force in the trafficking of nuclear supplies and materials from Russia to rogue nations and terrorist groups. In March 1993 the Chechens secured an unknown quantity of highly enriched uranium from Kazakhstan. More than six kilograms were transported from Grozny, the capital of Chechnya, to Istanbul, Turkey. Six months later, Interpol officials collared four Turkish businessmen and four agents of Iran's secret service and confiscated 2.5 kilos of the uranium.[11] The remaining kilos had slipped through the proverbial cracks. In September 1993 the Chechen rebels stunned the world by the display of SS-20 missiles during a military parade in Grozny. The missiles, stolen from some Soviet arsenal, possessed a range of 9,500 kilometers and the capability to launch a nuclear warhead.[12]

At the same time, the Chechen rebels began using radioactive isotopes, which had become available at rock-bottom prices, to commit acts of murder. The first victim was Vladimir Kaplun, the owner of a meat packing plant in Moscow. The Chechens planted gamma ray–emitting pellets in Kaplun's office. The businessman was dead within a matter of weeks.[13] At least half a dozen similar cases were reported in the next three years.

In November 1994 Dzokhar Dudayev, the leader of the Chechen rebels, petitioned the United Nations to dispatch troops to protect the weapons of mass destruction within the Chechen arsenal.[14] Dudayev's request was taken as an attempt at grandstanding. The UN officials, along with the international press, scoffed at the notion that a group of Muslim dissidents from the backwater Russian province of Chechnya could possess, let alone maintain, nuclear weapons. This forced the Chechens to make a vivid display that would prove to the world that they possessed nuclear capability. On November 23, 1995, Chechen commander Shamil Basayev directed a television crew to a radiological bomb that had been planted in Izmailovsky Park near Moscow.[15] The bomb was made of cesium-137 and, if detonated, would have killed hundreds in a matter of minutes and contaminated thousands more. The incident represented the first case of a dirty nuke to be shown as a weapon of terror.

Shortly after this incident, Dudayev notified the US State Department that he possessed tactical nuclear suitcase bombs. The Chechen leader said that he was willing to sell these weapons to rogue states or terrorist agencies, such as al Qaeda, if the United States failed to recognize Chechnya's independence from Russia. Dudayev's claim was supported by officials from the National Intelligence Council, an umbrella organization for the US analytical community. The officials informed a congressional committee that weapons-grade and weapons-usable nuclear materials, including SADMs (small atomic demolition munitions), had been stolen from Soviet stockpiles. "Of these thefts," the officials said, "we assess that undetected smuggling has occurred, although we do not know the extent or the magnitude."[16]

In January 1996 the Center for Nonproliferation Studies at the Monterey Institute of International Studies received information from a "senior advisor to Boris Yeltsin" that an unspecified number of small nuclear bombs had been manufactured for the KGB and had never appeared on a list of the Soviet nuclear inventory.[17]

In May 1997, during a closed door session with a US congressional delegation, former Russian Security Council secretary Alexander Lebed said that more than eighty-four SADMs had disappeared from Russian arsenals and could be in the hands of Muslim extremists. General Lebed said that he was able to confirm the production of 132 small nukes but could only account for 48. When asked the whereabouts of the missing nukes, Lebed replied: "I have no idea."[18] He went on to say that he had no idea how many small nukes had been produced by the Soviet Union during the cold war. Lebed also said that some of these small nukes, with an explosive yield of one kiloton of TNT, had been refined so that they could be carried by one person in a case measuring sixty by forty by twenty centimeters (twenty-four by sixteen by eight inches) and represented "ideal weapons for nuclear terror."[19]

General Lebed repeated these charges in an interview with the CBS newsmagazine *60 Minutes* that was aired on September 7, 1997. In the interview, he added that the SADMs were designed to be used in sabotage operations behind enemy lines and lacked the electronic combination locks that had been built into other Soviet nuclear weapons.[20] The following is a sample of the exchange between CBS correspondent Steve Kroft and General Lebed over the matter of the missing nukes:

KROFT: Are you confident that all of the weapons are secure and accounted for?

LEBED: Not at all. Not at all.

KROFT: How easy would it be to steal one?

LEBED: It's suitcase-sized.

KROFT: You could put it in a suitcase and carry it off?

LEBED: It is made in the size of a suitcase. It is a suitcase, actually. You could carry it. You could put it in another suitcase if you want to.

KROFT: But it's already a suitcase.

LEBED: Yes.

KROFT: I could walk down the streets of Moscow or Washington or New York, and people would think I'm carrying a suitcase?

LEBED: Yes, indeed.

KROFT: How easy would it be to detonate?

LEBED: It would take twenty, thirty minutes to prepare.

KROFT: But you don't need secret codes from the Kremlin or anything like that?

LEBED: No.

KROFT: You are saying there are a significant number that are missing and unaccounted for?

LEBED: Yes, there is. More than one hundred.

KROFT: Where are they?

LEBED: Somewhere in Georgia, somewhere in Ukraine, somewhere in the Baltic countries. Perhaps some are even outside those countries. One person is capable of triggering this nuclear weapon—one person.

KROFT: So you are saying these weapons are no longer under the control of the Russian military?

LEBED: I'm saying that more than one hundred weapons are not under the control of the armed forces of Russia. I don't know their location. I don't know whether they have been destroyed or whether they are stored or whether they've been sold or stolen. I don't know.[21]

Not all of the small Soviet nukes were designed to be carried around in an attaché case or a small suitcase by a single military operative or Soviet spy. Some were nuclear mines that were for use by engineering troops and deployed along the Soviet border, particularly the border with China. These nuclear mines were intended to create obstacles in the path of advancing armies by altering the landscape and producing high levels of radioactive contamination. At the end of the cold war, there were more than seven hundred of these mines in the Soviet stockpile. By 1997 Russian officials said that most, but not all, had been withdrawn to central storage facilities and eliminated.[22] Each device weighed less than two hundred pounds and produced an explosive yield of one to five kilotons.

Small nukes in the Soviet arsenal also included 120- to 150-pound atomic artillery shells for 152- and 155-millimeter howitzers. One on display at the public museum of Chelabinsk-70, Russia's nuclear design center, measures eighteen inches in length and six inches in diameter. It represents the world's smallest nuclear weapon.[23]

There were also backpack nukes, such as the Red Army's RA-155 and the Red Navy's RA-115-01 (for use underwater). These weighed less than seventy pounds and could be detonated within ten minutes. They were designed to destroy bridges, tunnels, airfields, communications facilities, and oil refineries.[24]

Some of the tactical nukes could fit nicely within the trunk of a car. Others would have to be transported in crates by truck or van. The smallest, with an explosive yield of .5 kilotons, could be used to obliterate a small town or village. The largest, with yields in excess of 15 kilotons, could be used to wipe out major metropolitan areas, such as New York and Los Angeles. In total, twenty-two thousand of these nukes had been manufactured by the Soviet army and became part of the standard equipment for Soviet forces stationed in Russia, East Germany, Hungary, Poland, Bulgaria, and Czechoslovakia.[25]

The controversy over Lebed's testimony came to center on his statement that small tactical nukes had been designed to be carried about in a suitcase with standard measurements. Russian prime minister Viktor Chernomyrdin dismissed this assertion as an "absolute absurdity" and said that all of his country's nukes had been accounted for and remained under strict control.[26] *Rossiyskaya gazeta*, Russia's official newspaper, went even further by stating that Lebed's charge represented the fantasy of a "diseased imagination."[27]

But Lebed's testimony received corroboration from Vladimir Denisov a former head of the Russian Security Council. Denisov, told a US congressional committee that tactical nukes meant to fit within a suitcase had been manufactured by the Soviet Union and that his office had received reports that several of the nukes had been missing and might have fallen into the hands of the Chechen separatists.[28]

Further corroboration came from Aleksey Yablokov, Environmental Advisor to Russian president Boris Yeltsin. In a letter published in the Moscow newspaper *Novaya gazeta* on September 22, 1997, Yablokov said that more than seven hundred nuclear suitcase bombs had been produced exclusively for the KGB. For this reason, he explained, the nuclear suitcase did not appear on the official inventory of Russia's nuclear weapons. He said that the present whereabouts of these weapons remained anyone's guess. He added that the small nukes would have required two major overhauls since the time they had been manufactured and it remained uncertain if such overhauls had been conducted.[29]

On October 6, 1997, Yeltsin signed a set of amendments to the Russian Federation Law on State Secrets that effectively classified all information about the nuclear suitcase bombs and military nuclear facilities in an effort to contain fears about the missing nukes.[30]

In the wake of these developments came news that the United States had developed and deployed weapons similar to those described by General Lebed throughout the cold war. One such weapon, based on the W-54 warhead, could fit neatly within the suitcase described by Lebed. It was only 27 inches long with a diameter of 11.2 inches and weighed less than sixty pounds. This weapon was crafted to fit in a backpack and be parachuted behind enemy lines for use by Navy SEALS. For this reason, it was commonly called a "rucksack bomb."[31]

The United States also manufactured thousands of "Davy Crockett" warheads that were deployed to strategic locations within the NATO alliance. These weapons, the "babies of American nukes," weighed less than fifty pounds and could be fired from 120- or 155-millimeter recoilless rifles. They produced an explosive yield of .25 kilotons, the equivalent of one hundred thousand sticks of dynamite.[32] The existence of these weapons provided substantiation for Lebed's claim, since there was no nuke in the United States arsenal that had not been replicated by the Soviets.

Most experts, including Carey Sublette of the Nuclear Weapon Archive, believe that the Russian nukes in question are 155-millimeter nuclear artillery shells that have been shortened by omitting the nonessential canonical ogive and fuse to fit within suitcases and even attaché cases.[33] To produce an explosive yield of 10 kilotons (enough to wipe out the five boroughs of New York City and much of Long Island), the weapons would have been fusion boosted by thin beryllium reflectors. Such devices would be based on a design approach called linear implosion. Linear implosion occurs when an elongated (oval-shaped) lower-density subcritical mass is compressed and deformed into a critical, higher-density spherical configuration by embedding it in a cylinder of explosives that are ignited at both ends. As the detonation progresses from both directions toward the middle, the fissile mass is squeezed into a supercritical mass.

In 2004 there were 4,000 tactical nuclear weapons, including the shells described above, in Russia; 3,300 in the United States; 400 in China; 200 in Israel; 60 to 80 in France; 60 in India, and 15 to 48 in Pakistan.[34]

While the controversy raged, the Clinton administration opted to ignore the claims of Chechen leader Dzhokhar Dudayev's that he possessed tactical nukes and was willing to sell them to the highest bidder. The administration refused to meet with him, let alone meet his demands for recognition of Chechnya's independence. Moreover, it even neglected to respond to the reports that the Chechen separatists had sold twenty nuclear suitcase bombs to Osama bin

Laden for $30 million in cash and two tons of choice Number Four heroin with a street value in excess of $700 million. Regarding this last issue, Rep. Curt Weldon said: "I am convinced back then our government didn't take the aggressive steps they should have taken to track down the stories. All during the 1990s they just brushed them aside."[35]

News of the sale of the nukes to bin Laden appeared on August 16, 1998, in the *London Times* and several weeks later in such publications as the *Jerusalem Report*, *Al-Watan Al-Arabi*, *Muslim Magazine*, and *Al-Majallah* (London's Saudi weekly).[36] For those within the intelligence community, the reports were hardly surprising. The Chechens rebels had close ties with al Qaeda. The two groups of radical Muslims had fought side by side in Azerbaijan from 1993 to 1994 to aid the Azeri *mujahadeen* in their struggle against Christian Armenia for control of the disputed Nagorno-Karabakh enclave.[37] When the war for independence broke out in Chechnya, at the end of the Azerbaijan-Armenia conflict, three hundred Afghan Arabs joined the six thousand Chechen guerrillas to ward off the invading Russian troops. In 1996 bin Laden established training camps for Chechen forces in Afghanistan, sent Amir Khattab and nine other al Qaeda commanders to oversee Chechen military operations, and provided a contribution in excess of $25 million as aid for the war effort.[38] In exchange for such assistance, the Chechens would have been pleased to sell the twenty suitcase nukes to the great emir. Indeed, they would have been hard pressed to find a more suitable buyer.

Because the Clinton administration opted to turn a deaf ear to accounts of the sale of the suitcase nukes to bin Laden, members of the national media came to assume that the reports must lack substance. Terri Whitcraft, an Emmy Award–winning producer at ABC, decided that the topic of nuclear terrorism was not a worthy subject for a news documentary because "visible proof" of this transaction could not be produced.[39] The network opted instead to air a report on Unidentified Flying Objects, hosted by Peter Jennings.

Yet the report had been substantiated by US, British, Russian, Israeli, Pakistani, and Saudi intelligence. It was even upheld by the United Nations. In 2004 Hans Blix, former Director General of the International Atomic Energy Agency (IAEA), told his colleagues at this world body that the accounts regarding the sale of twenty nuclear suitcase bombs to al Qaeda were accurate.[40] Dr. Blix reportedly made this announcement after meeting with Russian officials who had investigated the theft of the weapons and after speaking to Chechen leaders who had witnessed the transaction.

The accuracy of the story was also supported by the international press, including such news outlets as BBC, the *London Times, Al-Watan Al-Arabi,* and *Al-Majallah.* The celebrated Pakistani journalist Hamid Mir maintained not only that the story of the sale of the suitcase nukes was true but also that he had visited laboratories in Afghanistan, where al Qaeda scientists and technicians worked to maintain and upgrade the weapons.[41]

In 1998 Yossef Bodansky, Chairman of the Congressional Task Force on Terrorism and Unconventional Warfare in Washington, DC, told a congressional committee: "There is no longer much doubt that bin Laden has succeeded in his quest for nuclear bombs. The Russians believe he [bin Laden] has a handful [of nuclear weapons], the Saudi intelligence services are very conservative, perhaps they are friendly to the United States, believe he has in the neighborhood of twenty. As far as the acquisition and obtaining [of such weapons], there's the multiple sources of that, dealing with the actual purchase of the suitcase bombs. He [bin Laden] has a collection of individuals knowledgeable in activating the bombs and he is looking for and recruiting former Soviet Special Forces in learning how to operate the bombs behind enemy lines."

Asked about the immediacy of the threat, Bodansky said: "We don't have any indication that they are going to use it [the arsenal of suitcase bombs] tomorrow or any other day. But they have the capacity; they have the legitimate authorization; they have the logic for using it. So, one does not go into the tremendous amount of expenditures, effort, investment in human beings, and in human resources, to have something that will be kept in storage for a rainy day."[42]

In 1998 Shaykh Hisham Kabbani, chairman of the Islamic Supreme Council on America, testified before a committee of the US Department of State that Osama bin Laden had purchased the suitcase nukes from the Chechens; that many of these weapons had arrived in the United States; and that more than five thousand al Qaeda operatives were being trained for the American Hiroshima.[43] In the wake of September 11, 2001, Kabbani restated these claims to members of the press.[44]

Sources close to Tom Ridge, former director of Homeland Security, maintain that he shares Kabbani's belief that al Qaeda not only has secured the small nukes from the Chechens but has managed to smuggle them into the country.[45]

Further confirmation concerning the nuclear suitcases came with an alarm. On October 11, 2001, George Tenet, then director of the Central Intelligence Agency, met with President Bush to convey the news that at least two suitcase nukes had reached al Qaeda operatives within the United States. Each suitcase weighed between fifty and eighty kilograms (approximately 110 to 176 pounds) and contained enough fissionable plutonium and uranium to produce an explosive yield in excess of two kilotons. One suitcase bore the serial number 9999 and the Russian manufacturing date of 1988.[46] The design of the weapons, Tenet told the president, is simple. The plutonium and uranium are kept in separate compartments that are linked to a triggering mechanism that can be activated by a clock or a call from a cell phone.[47]

The news sent the president "through the roof," prompting him to order his national security team to give nuclear terrorism priority over every other threat to America.[48] It further caused the president to activate nuclear contingency plans—the first such measure since the dawning of the cold war—for the installation of underground bunkers away from major metropolitan areas so that a cadre of federal managers could proceed with the business of government if and when the nuclear attacks occurred.[49]

To reiterate the seriousness of the situation and the reality of the threat, Mr. Tenet said: "The threat environment we face is as bad as it was before September 11. It is serious. They have reconstituted. They are coming after us."[50]

Confirmation also came straight from the horse's mouth. To inform the world of his possession of the weapons, bin Laden issued the following statement called "The Nuclear Bomb of Islam": "It is the duty of Muslims to prepare as much force as possible to terrorize the enemies of God."[51] In a December 1999 interview with journalists from *Time* magazine, Bin Laden let it be known in an oblique way that he possessed nuclear weapons. This can be discerned from the following exchange:

> TIME: The U.S. says you are trying to acquire chemical and nuclear weapons. How would you use these?

> Bin Laden: Acquiring weapons for the defense of Muslims is a religious duty. If I have indeed acquired these weapons, then I thank God for enabling me to do so. And if I seek to acquire these weapons, I am carrying out a duty. It would be a sin for Muslims not to try to possess the weapons that would prevent the infidels from inflicting harm on Muslims.[52]

Several weeks later, when asked by John Miller of ABC News if he was seeking to obtain chemical and nuclear weapons, bin Laden said: "If I seek to acquire such weapons, this is a religious duty. How we use them is up to us."[53]

Moreover, as noted earlier, bin Laden let it be known during his interview with Hamid Mir in November 2001 that he possessed the nukes and was prepared to use them. He said: "I wish to declare that if America used chemical or nuclear weapons against us, then we may retort with chemical and nuclear weapons. We have the weapons as deterrent." When Mir asked the terrorist leader where he had obtained such weapons, bin Laden said: "Go on to the next question." As the interview continued, bin Laden told Mir that it had been relatively easy for al Qaeda to obtain the nukes. "It is not difficult, not if you have contacts in Russia with other militant groups. They are available for $10 million and $20 million." At this stage in the exchange, Ayman al-Zawahiri, bin Laden's chief strategist, interjected: "If you go to BBC reports, you will find that thirty nuclear weapons are missing from Russia's nuclear arsenal." Al-

Zawahiri smiled and added: "We have links with Russia's underworld channels."[54]

Bin Laden made more additions to his nuclear arsenal. In 1998 he purchased twenty nuclear warheads from Kazakstan, Turkmenistan, Russia, and the Ukraine. At the al Qaeda laboratories, scientists removed the active uranium and plutonium so that they could be processed and placed within backpack-sized nukes for easier transportation and less chance of detection.[55]

This account, too, had been ignored by the national media even when empirical confirmation was provided by a disturbing incident. In October 2001 the Mossad (Israeli intelligence) arrested an al Qaeda operative as he attempted to enter Israel through Palestinian territories at a border checkpoint in Ramallah.[56] Concerning the arrest, an Israeli official said: "There was only one individual involved. He was from Pakistan."[57] First reports of the incident in the international press spoke of the detection of radioactive material in the backpack, causing journalists to assume that the terrorist must have been carrying a radiological bomb. But, after reinterviewing the sources, United Press International and other reliable news sources came to the conclusion that the device was a tactical nuclear weapon. Israeli intelligence refused to issue an official comment but did confirm that the device was a complete weapon that required no assembly and could have been detonated by the operative. CIA officials later concluded that the backpack represented a plutonium-implosion bomb and not a "dirty nuke."[58] The proof positive had been provided. But, in the heyday of news carnivals about Michael Jackson, Martha Stewart, and the "runaway bride," few were paying attention.

CHAPTER SIX

The Five-Year Intermission

If you have $30 million, go to the black market in central Asia, contact any disgruntled Soviet scientist, and a lot of . . . smart briefcase bombs are available. They have contacted us, we sent our people to Moscow, to Tashkent, to other central Asian states and they negotiated, and we purchased some suitcase bombs.
—Ayman al-Zawahiri, interview with Hamid Mir, 2002

Since the advent of the Nuclear Age, everything has changed save our modes of thinking, and we thus drift toward unparalleled catastrophe.
—Albert Einstein, 1949

Lf the sale of the suitcase nukes actually took place in 1996, why hasn't the American Hiroshima already occurred? In the months that followed 9/11, the American people were filled with dread that the next attack would occur in the immediate future. It was expected to occur at the Super Bowl, the Academy Awards, the presidential conventions, the World Series, or on the Fourth of July. But nothing happened. Life in the US returned to normal. Flags no longer waved from most front porches; the terror level no longer rose and fell; church attendance returned to normal (with less than 30 percent of the members sitting in the pews); and books on al Qaeda and bin Laden were moved to the clearance tables of the nation's book chains. Although terrorist activities continued to escalate throughout the world since the launching of Operation Enduring Freedom, many Americans came to believe that the threat had been contained by the coalition forces.

The delay caused some to speculate that the bombs must have been duds. Lt. Gen. Igor Volynkin, head of the Russian ministry's Twelfth Main Directorate, the agency responsible for the storage and security of nuclear weapons, argued that al Qaeda obviously lacks the expertise to maintain the suitcase bombs. He said that the weapons must be disassembled every three months so that the nuclear cores can be recharged.[1]

American physicists at the Center for Nonproliferation Studies dispute this claim. They argue that the suitcase nukes, most likely, are uranium or plutonium devices that have been boosted with tritium. The tritium would compensate for the required amount of conventional explosive to compress the fissile core in the compact device. Neither the uranium nor the plutonium would need frequent maintenance. Even the tritium, with a half-life of 12.3 years, would not require constant replenishing.[2]

Nonetheless, the physicists agree, the weapons would require considerable technical attention. Most notably, the triggers that emit large quantities of neutrons at high speeds decay rapidly and have short half-lives—most would become useless in less than four months. The nuclear cores also are subject to decay and over the course of several years would fall below the critical mass threshold. Though the shells that encase the cores are the most durable parts of the weapon, they, too, are subject to contamination. Therefore, without proper care, the detonation of these devices would result in a fizzle rather than a boom.

The belief that bin Laden simply purchased these weapons for millions of dollars and stored them within his cave without concern for maintenance has its basis in the erroneous and prejudicial notion that he is a backward Bedouin warrior without knowledge of sophisticated weaponry, rather than a highly trained engineer and one of the most gifted military tacticians in the annals of modern history.

Bin Laden has been extremely mindful of proper maintenance. As soon as he obtained the weapons, he paid an amount estimated to be between $60 and $100 million for the assistance of nuclear scientists from Russia, China, and Pakistan.[3] From 1996 to 2001, bin Laden also kept a score of SPETSNAZ (Soviet special forces) technicians from the former Soviet Union on his payroll. These technicians had been trained to open and operate the weapons in order to prevent any unauthorized use.[4] To simplify the process of activation, the scientists and technicians came up with a way of hot-wiring the small nukes to the bodies of Muslim agents who long for immediate martyrdom and immediate elevation to seventh heaven.[5] The work on the weapons, for the most part, was conducted within laboratories that had been created within deep tunnels in the Khowst area and in the labyrinthine caves of Kandahar.[6]

Most speculation about the delay further fails to come to terms with the fact that one of bin Laden's defining characteristics is patience. He started plotting the 1998 bombings of the US embassies in Kenya and Tanzania when he was in Sudan in 1993; the attack on USS *Cole* was more than two years in the making; and eight years passed between the first attack on the World Trade Center and the second.[7] At the al Qaeda training camps, recruits were instructed to repeat this *surah* throughout the day: "I will be patient until Patience is outworn by patience."[8] The next attack, according to al Qaeda defectors and informants, will take place simultaneously at various sites throughout the country. Designated targets include New York, Boston, Philadelphia, Miami, Chicago, Washington, DC, Houston, Las Vegas, Los Angeles, and Valdez, Alaska, where the tankers are filled with oil from the Trans-Alaska pipeline. To orchestrate such an incredible event requires not only the shipment of the nukes into the United States but also the establishment of cells, the training of sleeper agents, the selection of sites, and the preparation of the weapons without detection from federal, state, or local law enforcement officials. Unlike 9/11, which cost less than $350,000,[9] this event already has cost a king's ransom and bin Laden will not waste the billions in expenditures, the years of planning, and his coveted "crown jewels" on an attack that is ill planned, poorly timed, and carelessly coordinated.

Another reason for the delay may reside in al Qaeda's attempt to retrieve the tactical nukes that were forward-deployed by the Soviets to the United States during the cold war. These weapons were buried at various remote sites throughout the country for recovery by Soviet agents when the long-anticipated war showdown would occur between the two superpowers. Many were suitcase devices. When the cold war turned hot, the weapons were to be uncovered and used to blow up dams, power stations, telecommunications centers, and landing strips for Air Force One.[10] "There is no doubt that the Soviets stored material in this country," Curt Weldon (R-PA), chairman of the House Armed Services Subcommittee on Military Research, said in 1999: "The question is what and where."[11]

In attempting to verify the stories about the buried nukes, FBI director Louis Freeh ordered a team of nuclear technicians to excavate several areas around Brainerd, Minnesota.[12]

Diggings began at additional sites soon after KGB defector Visili Mitrokhin, who served as a chief archivist for the agency, informed British intelligence that secret stockpiles of tactical nukes were buried in upstate New York, California, Texas, and Montana. On January 4, 2000, he repeated these charges before a US House Committee on Government Reform.[13]

Mitrakhin's information was supported by Col. Stanislav Lunev, the highest-ranking military spy to defect from the Soviet Union and the leading confidential source for the CIA on Russia's nuclear arsenal. Lunev told the same congressional committee that nuclear suitcases, indeed, had been buried throughout the United States, although he could not pinpoint the exact locations. Such information remains secret, Lunev said, because Russian military leaders continue to believe that a nuclear conflict between Russian and the United States is "inevitable." He concluded his testimony by saying: "And just now what we are talking about, location of technical nuclear devices, these places we have selected extremely carefully for a long, long period of time, and to believe it is possible to find these places just like that without using extremely, extremely large resources of the country, I don't think that it would be realistic until the Russian government, which still has the keys to these places, will disclose their locations."[14]

During the hearings, Belgian officials testified that they had found three secret depots replete with radio sets and tactical nukes that had been buried in Belgium by the Soviets during the 1960s.[15]

The number of nukes that remain buried in the United States is anyone's guess. But Soviet scientists produced more than seven hundred nuclear suitcases during the 1960s and 1970s and hundreds more during the 1980s. These weapons were placed under the care of SPETZNAZ technicians for deployment and detonation. Many of these scientists and technicians, during the 1990s, were sought out and employed by bin Laden.

There's a final reason for the delay. While most of the nukes in the al Qaeda arsenal are lightweight (less than sixty pounds) and easily portable in suitcases and backpacks, others are crude, cumbersome devices that weigh between one thousand and two thousand pounds. Such weapons would have to be shipped in cargo containers. And the containers would have to be lined with lead shielding to block the gamma rays in order to prevent radiation detection at ports of entry.

These weapons are called "gun bombs." Like a rifle, they use a conventional explosive charge to fire a bullet. Only in this case, the bullet is a lump of enriched uranium that slams into another piece of enriched uranium at the end of the barrel. The impact compresses the two pieces of uranium into a supercritical mass that sets off a nuclear chain reaction.[16] The design is simple and would be very attractive to bin Laden. It was the design of the nuclear bomb that leveled Hiroshima in 1945.

Casting the bomb's uranium into two pieces would be an easy task for a trained nuclear technician. A more daunting task would be designing the cannon to fire the uranium bullet. The terrorists would have to fashion their own cannon or, better yet, acquire a standard piece of military hardware like a Howitzer. The choice would depend upon the quality of the uranium. The more highly enriched the uranium, the less powerful the cannon that would be needed to produce the chain reaction.[17]

How difficult would it be to create such a weapon? Senator Joseph F. Biden, the highest-ranking Democrat on the Foreign Relations Committee, decided to find an answer. "I gathered the heads of all the national laboratories and asked them a simple question," Biden recalled in January 2004. "I said, 'I would like you to go back to your laboratory and try to assume for a moment you are a relatively informed terrorist group with access to some nuclear scientists. Could you build, off the shelf, a nuclear device? Not a dirty bomb, but something that would start a nuclear reaction—an atomic bomb.' They came back several months later and said, 'We built one.' They put it in a room and explained how—literally, off the shelf, without doing anything illegal—they actually constructed the device."[18]

"Off the shelf" works that provide step-by-step instructions for the creation of an atomic bomb are available at book stores throughout the country. Two such works—*The Los Alamos Primer* and *Atomic Energy for Military Purposes*—can be obtained from Amazon.com for a combined cost of $40.76 plus shipping.[19]

In 1979 James R. Schlesinger, secretary of energy for the Carter administration, sought a federal injunction to halt the publication of "The H-Bomb Secret" in the *Progressive* magazine. Schlesinger's action represented only the second time in US history that the government attempted to block the publication of a work because it represented a threat to national security. The first attempt came during the Nixon administration with Daniel Ellsberg's *The Pentagon Papers*. "The H-Bomb Secret" provided a description of the construction of a hydrogen bomb in such detail that it would have been classified top secret if it had been a government document.[20] After a six-month court battle, the government lost. "The H-Bomb Secret" was published in the *Progressive*, exactly as written and with a set of schematics for anyone interested in building a thermonuclear weapon. The article is now available free of charge on the Internet.

Bin Laden's attempt to manufacture his own nukes is evidenced by his expenditure of millions for the necessary raw material (highly enriched uranium and/or plutonium). His quest for such material began in 1993, when he attempted to purchase highly enriched uranium (HEU) from South Africa—a country that had a nuclear program under its previous government.[21]

Relatively new to the free-for-all thieving of the post-Soviet republics, bin Laden was fleeced on more than one occasion by wily con artists. He shelled out $200 million for a supposedly essential ingredient for a secret weapon. The ingredient turned out to be low-level nuclear waste that could be used only for a radiological device.[22]

But, more often than not, the search was successful, culminating in the purchase of more than twenty kilos of uranium-236 from Semion Mogilevich, a Ukrainian arms dealer. For one delivery of twelve to fifteen kilos, Mogilevich received a payment of $70 million.[23] The uranium had been enriched to 85 percent—far above the standards of enrichment for weapons-grade material.[24] On March 3, 2000, an Egyptian named Ibrahim Abd and several Congolese opposition soldiers sold two bars of enriched uranium-138 to al Qaeda agents at a meeting place in Hamburg, Germany.[25] Bin Laden also obtained enough radioactive material from black market sources in Russia, China, Kazakhstan, and the Ukraine to build a host of dirty nukes. He was gracious enough to provide proof of this fact by leaving behind a lead-lined canister for US troops to find within a cave in Kandahar.[26]

But the best reason for believing that bin Laden manufactured his own crude nuclear weapons from the highly enriched uranium and plutonium that he managed to obtain from the black market comes from his ties to Pakistan—the land of the Taliban and the home base of the *mujahadeen*. Pakistan, with a population of 145 million, remains the largest Islamic nation on earth—97 percent of the people are Muslims, of which 77 percent are Sunnis and 20 percent Shiites. The country is home to thousands of radical mosques and *madrassahs* along with a host of terrorist organizations, including Jamaat-e-Islami, Jaish Mohammed, Hezb-ul-Mujahadeen,

Lashkar-e-Toiba, al-Hadith, Harakat ul-Ansar, and al-Badar. Bin Laden's image remains omnipresent. It looms in murals on the sides of buildings, on posters nailed to utility poles, in the windows of stores and shops, and on the mud-brick and wattle stalls that line the marketplaces. He has become the great hero of the Pakistani people—second only to Dr. Abdul Qadeer Khan, the father of the Islamic bomb. The two are united. Bin Laden has met with Khan and various members of his nuclear research facility on the outskirts of Islamabad. It is safe to assume that the meetings were not held to discuss the latest game of *buzkashi* or the weather conditions in Balochistan.

Dr. Khan represents the greatest nuclear proliferator in history. He has provided centrifuge technology for uranium enrichment and blueprints for nuclear weapons to such rogue nations as North Korea, Iran, Libya, and parts of Brazil. If Khan remained willing to share such knowledge with infidels such as Kim Jong II and apostates such as Muammar Gadhafi, why would he not supply the same information to a true believer with deep pockets such as Osama bin Laden?

CHAPTER SEVEN

Enter Dr. Evil

If a nuclear weapon destroys the U.S. Capitol in coming years, it will probably be based in part on Pakistani technology.
  —Nicholas D. Kristof, *New York Times*, September 27, 2004

We have the nuclear capability that can destroy Madras; surely the same missile can do the same to Tel Aviv. Washington cannot stop Muslim suicidal attacks. The Taliban are still alive and, along with "friends," they will continue the holy war against the United States
  —General Hamid Gul, former head of Pakistan's ISI
  (Inter-Services Intelligence), 2002

I have a nightmare that the spread of enriched uranium and nuclear material could result in the operation of a small enriched facility in a place like northern Afghanistan. Who knows? It's not hard for a nonstate to hide, especially if a state is in collusion with it. Some of these nonstate groups are very sophisticated.
  —Mohamed el-Baradei, director-general of the International
  Atomic Energy Agency (IAEA), 2004

Dr. Abdul Qadeer Khan, a German-trained metallurgist, began his career by stealing plans for enriching uranium to weapons-grade strength in gas centrifuges from Urenco, a top-secret atomic energy plant in the Netherlands. He had been employed there as a technician. Khan fled to his native Pakistan, where he set up his own nuclear testing facility—the A. Q. Khan Research Laboratories—near Islamabad. For the next twenty-eight years, he worked to build "the Islamic bomb" with the assistance of Chinese technicians and scientists.[1] His efforts were subsidized by wealthy Saudi Arabian businessmen and the administration of Pakistani prime minister Benazir Bhutto. Because Pakistan was providing support for the CIA's covert actions in Afghanistan to ward off the Soviet invasion, the United States took no steps to thwart the clandestine efforts of Khan to develop his weapons.[2]

On May 28, 1998, thanks to Khan's efforts and thievery, Pakistan officially joined the elite nuclear club of nations, including the United States, Russia, China, Great Britain, France, India, and Israel with the successful testing of five atomic bombs beneath the scorched hills of the Baluchistan desert. This day has become Yaum-e-Takbeer, a national holiday, in which eerie light shows recreating the giant mushroom in the sky are presented throughout Pakistan.[3] In towns and villages, celebrants worship the radioactive fragments from the blast as religious artifacts that can bestow spiritual powers upon families and country.[4]

By stabilizing the nuclear threat from India, Dr. Khan emerged as a national hero in Pakistan, where his birthday is sanctified in mosques.[5] He is the only Pakistani to have received the highest civilian award of "Nishan-i-Imtiaz" twice—in 1996 and 1998.[6] "His stature is so elevated and protected by the Pakistani government," writes Rajesh Kumar Mishra of the South Asia Analysis Group, "that he has been epitomized as larger than the 'nuclear image' of Pakistan. Anything said or done against this scientist is considered anti-Pakistan, anti-Islam, and intolerable."[7]

In the United States and western Europe, however, Dr. Khan

became viewed as a rogue scientist, a nuclear madman, and an individual far more dangerous than Osama bin Laden or Ayman al-Zawahiri. Several years before the mushroom cloud appeared over the Chagai Hills, Dr. Khan, at the request of Prime Minister Bhutto, engaged in discussions with North Korean officials for the purchase of twelve to fourteen Nodong ballistic missiles. A deal was struck. In exchange for the six-hundred-mile-range, nuclear-capable missiles that would be modified to produce Pakistan's Ghaudi missiles, Dr. Khan provided North Korea with the Dutch blueprints he had stolen for the enrichment of uranium. The deal enabled North Korea to emerge as the most formidable nation in President Bush's infamous "axis of evil."[8]

In 2001, Dr. Khan met with Iranian president Mohammad Khatami and offered to place the country's nuclear weapons program on the fast track by providing Iran with the same pilfered blueprints that he had used in Pakistan and had delivered to North Korea. Within a matter of months, he reportedly implemented a program for the enrichment of uranium at the Bushehr Nuclear Power Plant and oversaw the construction of nuclear facilities in Arak and Natanz.[9] Iran, according to most estimates, is expected to become a full-fledged member of the nuclear club of nations by 2006.[10]

Dr. Khan's sale of nuclear blueprints to Iran likely served as a basis for President Bush's decision to launch an invasion of Iraq in March 2003. Without toppling Saddam Hussein, propping up a friendly regime, and establishing strategically located military bases throughout Iraq, the United States could not hope to contain the growing nuclear threat from Iran.[11]

Dr. Khan offered his blueprints to other Islamic countries. In a letter to Saddam Hussein, before the dictator was toppled from power, the Pakistani scientist made an offer "to establish a project to enrich uranium and manufacture a nuclear weapon for Iraq."[12] He made a similar proposal to Libyan president Muammar Gadhafi.[13] At that point, Gadhafi accepted and Dr. Khan became the leading supplier of nuclear know-how to yet another Muslim country.[14] Gadhafi's attempt to join the nuclear club became thwarted when

Italian officials intercepted a shipment of centrifuge parts to Libya in October 2003.[15] Two months later, the Libyan dictator announced that Libya would abandon its nuclear program and destroy its arsenal of weapons of mass destruction.[16] This announcement, however, did not deter Khan, who went on to sell his centrifuge technology to Saudi Arabia, Sudan, Nigeria, and Brazil.

As the godfather of nuclear proliferation, Dr. Khan employed SMB Computers, an UAE company in Dubai, to serve as the go-between agency for the transfer of nuclear weapons components to the rogue nations.[17] He also established a factory in Malaysia for the mass production of centrifuges.[18]

By 2002 business at the Malaysian factory was booming. Tens of thousands of centrifuge parts were produced for the Pyongyang government, bringing North Korea to the threshold of unlimited bomb production.[19] Similar shipments had been prepared for Iran, and orders were backlogged from countries throughout the world.

When news of these shipments reached the White House, President George W. Bush insisted that Pyongyang and Tehran cease and desist from producing highly enriched uranium and plutonium and from producing nuclear warheads. But his national security team failed to come up with a policy to make the two countries submit to this demand. By 2005, the stalemate had left three secret overtures from Tehran unanswered and a presidential directive on Iran unsigned after thirty-one months of attempts by the president's staff to produce a persuasive draft. A similar impasse over North Korea left the Bush administration with a policy that one official described as "no carrot, no stick, and no talk."[20]

Khan also nurtured a close relationship with China. He provided Beijing with his centrifuge technology to replace the country's outdated enrichment techniques of uranium enrichment in exchange for advanced designs of implosion-type nuclear bombs. This deal was great for Khan, who turned around and sold the designs to several of his long-standing customers, including North Korea and Iran.[21]

In addition to his pivotal place in creating the so-called axis of evil, Dr. Khan remained a proponent of radical Islam, professing his opinion that "all Western countries are not only the enemies of Pakistan, but, in fact, of Islam."[22] He appeared in rallies and at conclaves of Lashkar-e-Toiba ("the Army of the Pure"), the officially banned terrorist arm of Markaz Dawa-Wal-Irshad, a Sunni organization of the Wahhabi sects in Pakistan.[23] As noted earlier, Wahhabi sects follow the radical teachings of eighteenth-century emir Abdul al-Wahhab. Bin Laden, al-Zawahiri, and all high-ranking al Qaeda officials are Wahhabists.

Lashkar-e-Toiba specialized in the mass murder of Hindus and was responsible for several horrific terrorist attacks, including the slaughter of twenty-three people in Wandhama on January 23, 1988; the cold-blooded slayings of twenty-five members of a wedding party in Doda on June 19, 1998; the Chattisinghpora massacre of March 20, 2000, in which thirty-five men, women, and children were hacked to pieces by machetes; and the November 24, 2002, attack on two Hindu shrines in Jammu that left thirteen people dead and forty-five injured.[24] In 2003, Lashkar-e-Toiba changed its name to Jamaat-ud-Dawa and became the coordinating agency for bin Laden networks throughout the world.[25] On February 2, 2002, Abu Zubaydah, a member of al Qaeda's Majilis al Shura (Central Consultation Council), was found hiding within a Lashkar-e-Toiba safe house in Faisalabad. A key strategist of 9/11, Abu Zubaydah was the terrorist chiefly responsible for the instruction and dispatch of Richard Reid, the alleged "shoe bomb" who attempted to blow up an American Airlines plane in December 2001.[26] Within Zubaydah's address book were the names of several key al Qaeda operatives throughout Latin America.[27]

Throughout 2004 Hafiz Mohammed Saeed, the emir of Lashkar-e-Toiba (now Jamaat-ud-Dawa), barnstormed the provinces of Pakistan, calling for all believers to join in the *jihad* against the United States. In several speeches, Saeed maintained that Pakistan's nuclear weapons should be used to benefit all Islamic states. In addition, he claimed that his terrorist organization remained in control of "two nuclear missiles." These missiles, Saeed said, were being prepared to be used against the "enemies of Islam."[28]

At Lashkar-e-Toiba rallies and gatherings, Dr. Khan was often joined by other leading scientists from his research laboratory, including his close friend and colleague Dr. Sultan Bashiruddin Mahmood.[29]

On October 23, 2001, US military and intelligence officials discovered records of meetings between Dr. Sultan Bashiruddin Mahmood and al Qaeda officials, including bin Laden and al-Zawahiri, within the headquarters of Ummah Tameer E-Nau ("Islamic Reconstruction") in Kabul. This organization served as a front for al Qaeda by posing as a charitable organization "to serve the hungry and needy of Afghanistan."[30] The finding was so alarming that CIA director George Tenet boarded a midnight flight to Islamabad to investigate the matter.[31]

Dr. Sultan Bashiruddin Mahmood, the founder and president of Ummah Tameer E-Nau (UTN), represented the perfect person to provide nuclear materials and nuclear expertise to al Qaeda. He had spent more than twenty years working on the enrichment of uranium at the Khan Research Laboratories. Dr. Mahmood's expertise was so respected by his colleagues that in 1995 he was named chairman of the Pakistan Atomic Energy Commission. He also served as the executive director of the Khosab reactor in the Punjab region—a reactor that produces weapons-grade plutonium.

Along with his hands-on technical expertise with nuclear weapons, Dr. Mahmood, like other officials at the Khan Research Laboratories, possessed an extensive knowledge of black market sources for fissile materials throughout the Middle East and Asia. On several occasions, Mahmood said he wished to share this knowledge with the *mujahadeen* and all Muslim nations for the sake of (what he believed to be) "the last war between Islam and the infidels."[32]

"Mahmood was one of the nuclear hawks," said Rifaat Hussain, the chairman of defense and strategic studies at Quaid-i-Azzam University in Islamabad. "People say that he was a very capable scientist and a very capable engineer, but he had this totally crazy mind-set."[33]

Evidence of this craziness can be found in his writings. In *Mechanics of the Doomsday and Life after Death*, he argues that natural catastrophes occur in places where moral degradation becomes rampant.[34] In *Cosmology and Human Destiny*, he maintains that sunspots determine the course of human history, including World War II, the revolution against colonial government in India, the Soviet invasion of Eastern Europe, and the rise of the *mujahadeen*.[35] In this same work, Mahmood says that Pakistan's energy problems could be solved by harnessing the energy emitted by the *djinn*, the fairylike spirits who inhabit the world of Islam.[36]

In 1999, after urging nuclear proliferation, advancing the cause of Lashkar-e-Toiba, and making a string of bizarre statements, Dr. Mahmood was forced to step down from his positions as chairman of Pakistan's Atomic Energy Commission and executive director of the Khosab facility. He packed up his bags (including, many believe, nuclear designs and materials) and headed off to Kabul to establish his "charity."

Following the discovery of records of his meetings with bin Laden at UTN headquarters, Dr. Mahmood was placed under arrest by Pakistani Inter-Services Intelligence (ISI) agents and interrogated by the CIA about the records of his meetings.

At first, Dr. Mahmood denied that he had ever met bin Laden or made contact with al Qaeda officials. Later, he acknowledged that he had met with Mullah Omar, the supreme leader of the Taliban, but only to discuss the building of a flour mill in Kabul to provide bread for the masses. He continued to insist that he knew bin Laden only by reputation as a fellow humanitarian who "was helping in different places, renovating schools, opening orphan houses, and helping with the rehabilitation of widows."[37]

After several weeks on the hot seat, Dr. Mahmood finally admitted that he had met with Osama bin Laden, al-Zawahiri, and other al Qaeda officials in Kabul, including the fateful morning of September 11, 2001, to discuss the benefits of an atomic blast in an American city.[38] He said that bin Laden was most sincere in his efforts to create an American Hiroshima; that al Qaeda already possessed fissile materials from the Islamic Movement of Uzbekistan and other sources; and that members of the terrorist organization were actively engaged in speeding up the process of producing nukes for the *jihad*.[39] Dr. Mahmood, however, continued to insist that he had declined to provide nuclear materials and hands-on assistance to bin Laden for his Manhattan project. Upon making this denial, he was subjected to six lie-detector tests. He failed them all.[40]

As they continued to grill Dr. Mahmood, US officials discovered that bin Laden had held clandestine meetings with many other scientists from the A. Q. Khan Research Laboratories, including Dr. Chaudry Abdul Majid, the chief engineer of the Pakistan Atomic Energy Commission. After being threatened with seven years in prison under Pakistan's Official Secrets Act, Dr. Majid admitted that he had met with bin Laden and other al Qaeda officials to "discuss" the construction and maintenance of nuclear weapons.[41]

After months in Pakistani custody, Mahmood and Majid were quietly released. Knowing that Ummah Tameer E-Nau was a front for al Qaeda, the Bush administration placed the "charity" on its "terrorist list and designated Mahmood as a "global terrorist." The Pakistani government never placed the two scientists on trial, and they remain free men as of this writing.

Then there were the scientists from Khan's research facility who got away before US interrogators could collar them. Dr. Mohammad Ali Mukhtar and Dr. Sulieman Asad, nuclear engineers and close colleagues of Dr. Khan and Dr. Mahmood, managed to slip out of Pakistan and head to Myanmar. Pakistani officials said that the two had been dispatched by the government on an unspecified "research project." Mukhtar and Asad, who worked for many years at two of Pakistan's most secret nuclear weapons installations, had met with bin Laden and other Al Qaeda officials on numerous occasions.[42] In January 2002 the *Wall Street Journal* reported that Mukhtar and Asad were aiding Myanmar's efforts to build a ten-megawatt nuclear reactor.[43]

Other senior nuclear scientists from the Khan Research Laboratories escaped for unknown destinations. The list of such "absconders" includes the names of Muhammad Zubair, Murad Qasim, Tariq Mahmood, Saeed Akhther, Imtaz Baig, Waheed Nasir, Munawar Ismail, Shaheen Fareed, and Khalid Mahmood.[44]

Along with the statements of Mahmood and Majid, US investigators obtained documentary evidence from the offices of Ummah Tameer e-Nau that al Qaeda had succeeded in producing several nuclear weapons from highly enriched uranium and plutonium pellets that were the size of silver dollars. Of even greater concern was the evidence that at least one weapon had been shipped out of Afghanistan before 9/11. The weapon had been transported by al Qaeda agents to Karachi, where it reportedly was shipped to the United States in a cargo container.[45] There existed over eighteen million potential delivery vehicles that could be used to bring this nuclear weapon into the United States. That figure represents the number of cargo containers that arrive into the country every year. Of these containers, only 3 percent are inspected. Moreover, the bills of lading do not have to be produced until the containers reach their place of destination.[46] Despite the compelling urgency of this news to the American people, the report was not carried by the national media.

Telltale proof of Dr. Khan's personal involvement with bin Laden's Manhattan Project came on March 2, 2003, with the arrest of Khalid Shaikh Mohammed, al Qaeda's military operations chief, in Karachi. From Mohammad's laptop, US intelligence officials uncovered documentary evidence of meetings between Dr. Khan and Osama bin Laden within a safe house in Kabul.[47] This finding was supplemented by Khalid Mohammad's admission to CIA interrogators that bin Laden's goal was to create a "nuclear hell storm" within the United States and that the chain of command for this operation answered directly to bin Laden, al-Zawahiri, and a mysterious scientist called "Dr. X."[48]

US investigators also amassed evidence that agents of Pakistan's Inter-Services Intelligence (ISI) executed American reporter Daniel Pearl because the journalist had obtained confidential information concerning the clandestine meetings between Dr. Khan and bin Laden and on the trafficking of nuclear blueprints and materials from Khan's research laboratory near Islamabad to al Qaeda cells in the North-West Frontier Province of Pakistan.[49]

Under growing pressure from the United States, President Pervez Musharraf retired Dr. Khan as head of Pakistan's nuclear programs and appointed him as "special presidential advisor." Dr. Khan's demotion was decried by Nawaz Sharif, the former prime minister, as a hideous conspiracy designed to roll back the nuclear program and weaken the country's defenses.[50] President Musharraf attempted to impart a positive spin to his decision by saying: "Nations cannot afford to sit on their laurels. Success must be reinforced. New ideas and new blood must be injected. I always believed that the time to make a transition is when you are on top. I also believe that transition must be effected smoothly so that there is no dislocation of objectives. Giving any other color and meaning to my decision is unfair."[51]

On February 4, 2004, Dr. Khan issued a public statement in which he confessed that he had sold his centrifuge technology for the creation of highly enriched uranium to Libya, Iran, and North Korea. He expressed "the deepest sense of sorrow and anguish" that he had placed Pakistan's national security in jeopardy. "I have much to answer for it," he said.[52] Pakistan's Federal Cabinet and President Musharraf responded to Dr. Khan's statement by granting him a full act of pardon for selling state secrets and promoting nuclear proliferation. Musharraf said that Dr. Khan and the scientists who worked with him were simply motivated by "money."[53] The pardon, according to some observers, represented an attempt by the Musharraf government to appease Islamic extremists and senior Pakistani military officials who believed that Musharraf had become a traitor to the Muslim people by providing military support and assistance to the Bush administration.[54] The Pakistani president, however, agreed to limit Dr. Khan's travel to other countries and to have two military officials accompany the public hero on his trips to the various provinces of Pakistan.[55]

Throughout 2003 and 2004, additional information regarding Dr. Khan came to light, including the fact that he had provided to Iran not only the blueprints for centrifuge technology but also the designs for nuclear warheads.[56] This discovery was made by experts from the International Atomic Energy Agency (IAEA) who came upon blueprints for a ten-kiloton atomic bomb that Dr. Khan had delivered to Libya. Upon seeing the blueprints, the inspectors realized that they had come upon a level of danger that threatened global security. "This was the first time we had ever seen a loose copy of a bomb design that clearly worked," an IAEA official told the *New York Times,* "and the question was: Who else had it? The Iranians? The Syrians? Al Qaeda?"[57]

IAEA investigators further learned that Dr. Khan made business trips to eighteen countries for the purpose of buying nuclear materials, such as uranium ore, or selling atomic goods.[58] The list of Dr. Khan's suspected customers became expanded to include Egypt, Malaysia, Indonesia, Algeria, Kuwait, Myanmar, and Abu Dhabi.[59]

Despite the urgency of these findings, US intelligence officials were denied the right to question Dr. Khan about his nefarious activities. "It's an unbelievable story, how this administration has given Pakistan a pass on the worst case of proliferation in the past half century," said Jack Pritchard, who served as the US State Department's special envoy to North Korea. "We've given them a pass because of [Pakistani president] Musharraf's agreement to fight terrorism, and now there is some suggestion that the hunt for Osama is waning. And what have we learned from Khan? Nothing."[60]

Amazingly, more than a year after Dr. Khan's public confession, neither the US nor its allies possessed a full list of his customers. Musharraf acknowledged this by saying: "I can't say that we have unearthed everything that he's done, but I think we have unearthed most of what he's done."[61] In other words, Musharraf could offer no assurance that bin Laden did not number among Khan's customers.

Moreover, throughout the horrific affair, Musharraf continued to insist that his government was not involved with Khan and his asso-

ciates in providing nuclear secrets to rogue nations, let alone terrorist groups. "I am extremely positive neither the government nor
the military was involved," he said. "The Pakistan government had
carried out investigations and concluded that it was these individuals [Khan and associates] who carried out the proliferation of
nuclear technology."[62]

Yet the evidence shows otherwise. Throughout his travels to various countries, Dr. Khan remained in constant contact with the Pakistani Foreign Office. The Foreign Office, in turn, made sure that the
official diplomats in the various countries presented Khan with due
honors of protocol.[63] Moreover, Dr. Khan was always accompanied
by senior serving scientists of Pakistan's nuclear establishment who
could not have gone abroad and remained absent for extended
periods of time without the permission of Musharraf's government.[64]

And then there were the brochures—a glossy sales publication
from Khan's research laboratories that had been presented to rogue
nations and terrorists groups throughout the world. A photo of Dr.
Khan superimposed over a mushroom cloud appeared on the cover.
The contents consisted of a catalog of unique products and services
offered by Khan's network, including such items as "starter kits" for
uranium enrichment; P-1, P-2, and P-3 centrifuge designs; blueprints for nuclear warheads; and consulting services for assembly
and repair. The brochures bore the official seal of the Government
of Pakistan.[65]

Such discoveries forced many US military officials and weapons
experts to conclude that the greatest threat to national security
comes from the country's newest ally. Robert Gallucci, a former UN
weapons inspector and dean of Georgetown University's School of
Foreign Service, said: "Bad as it is with Iran, North Korea, and Libya
having nuclear-weapons material, the worst part is that they could
transfer it to a nonstate group. That's the biggest concern and the
scariest about all this—that Pakistan could work with the worst terrorist groups on earth to build nuclear weapons. The most dangerous country for the US now is Pakistan, and the second is Iran.
We haven't been this vulnerable since the British burned Washington in 1814."[66]

PART THREE

From Hell to South America

Has not there come to you the news of the overwhelming calamity?
(Some) faces on that day shall be downcast,
Laboring, toiling,
Entering into burning fire,
Made to drink from a boiling spring.
They shall have no food but of thorns,
Which will neither fatten nor avail against hunger.
(Other) faces on that day shall be happy,
Well-pleased because of their striving,
In a lofty garden,
Wherein you shall not hear vain talk.
Therein is a fountain flowing,
Therein are thrones raised high,
And drinking-cups ready placed,
And cushions set in a row,
And carpets spread out.
Will they not then consider the camels, how they are created?
And the heaven, how it is reared aloft,
And the mountains, how they are firmly fixed,
And the earth, how it is made a vast expanse?
Therefore, do remind, for you are only a reminder.
You are not a watcher over them;

But whoever turns back and disbelieves,
Allah will chastise him with the greatest chastisement.
Surely to Us is their turning back,
Then surely to Us is the taking of their account.

—"The Overwhelming Calamity," Koran 88

CHAPTER EIGHT

Welcome, Osama, to South America

The tri-border region of South America is the Hilton of Islamic extremism. It's been on the radar screen since the early '90s but no one's ever done anything about it.
> —Magnus Ranstorp, director of the Centre for the Study of
> Terrorism and Political Violence, University of St. Andrew,
> Fife, Scotland, 2003

In the tri-border area of Argentina, Brazil and Paraguay, Middle East terror organizations such as HAMAS and Hezbollah train terrorists and conduct fundraising activities in an area which has a growing population of Middle Eastern and South Asian immigrants. Funds raised in the tri-border area are sent directly to the Middle East to support the operation of these organizations, possibly even the planning and execution of ter-rorist acts. No doubt funds raised in the tri-border area have made it to the pockets of al Qaeda and Osama bin Laden. The FBI claims Islamic extremist cells linked with Hezbollah, Islamic Jihad and al Qaeda are operating in Paraguay, Uruguay and Ecuador.
> —Transcript of House Western Hemisphere
> Subcommittee Report, October 10, 2001

Hezbollah is the A-Team of Terrorism.
—Senator Bob Graham,
chairman of the US Intelligence Committee

I think you have an "atomic bomb" brewing between bin Laden, Hezbollah, and the Iranians. If these huge forces are married, either could set off the spark.
—Kenneth Katzman, terror analyst for the US Congress, 2003

If we don't want these facilities in Iran or North Korea, we shouldn't want them in Brazil.
—James E. Goodby, former US nuclear negotiator, 2004

At its announced capacity, Brazil's new facility at Resende will have the potential to produce enough U-235 to make five to six implosion-type warheads per year. By 2010, as capacity rises, it could make enough every year for 26 to 31, and by 2014 enough for 53 to 63.
—Liz Palmer and Gary Milhollin, *Science* magazine,
October 22, 2004

Thinking about an exotic place for your next vacation? Forget Fiji and Tahiti. Don't think twice about Christmas Island or Easter Island. Overlook such overstuffed stops as Bermuda and the Bahamas. Ignore such tourist traps as Barbados and Trinidad. Consider, instead, Iguaçu Falls, a remote location in South America that contains one of the world's most awesome wonders.

Iguaçu Falls (which was shown in the 1986 film *The Mission*) is situated in the meeting point between Argentina, Brazil, and Paraguay, an area known as the "Triangle." It consists of 273 cascades that are spread into a horseshoe shape over two sprawling miles of the Rio Iguaçu.

Accommodations are far from primitive. There are several five-star hotels on the Brazilian side, including the Cataratas del Iguaçu and the Sheraton International, which offer balconies that overlook the Devil's Throat (*garganta del diablo*), where fourteen falls drop 350 feet with such force that there is also a 100-foot cloud of spray overhead. Upon seeing this sight for the first time, Eleanor Roosevelt exclaimed: "Poor Niagara!"

With the help of a local guide, you can spend an afternoon wandering through the lush primordial rain forest with its giant and strangulating trees, natural palmettos, giant ferns, bromeliads, begonias, and orchids. Here you will find yourself surrounded by tropdurus lizards, capuchin monkeys, plumed macaws, and brightly colored parrots. You will be able to spot such rare and exotic creatures as the aardvark-like *coatis* that feast upon the mounds of red army ants, the crocodile-like *caimans* that lurk in the steamy ponds, the pumas that prowl the dense jungle in search of their next meal, and, of course, the pretty pink tarantulas.

Naturally, it is quite dangerous to stray from the jungle path. You could sink into a pit of quicksand, stumble upon a hungry jaguar, or fall prey to a pack of wild dogs. Even worse, you might wander into one of the three cities a few miles from the picturesque Iguaçu Falls without a semiautomatic machine gun strapped to your side.

The central city within the "Triangle" is Ciudad Del Este ("City of the East") in Paraguay. It represents a major base for the shipment of Bolivian cocaine via hidden landing strips in the rain forests. More than $12 billion a year passes through this remote city in the midst of the South American jungle. Within the stores and galleries of this dingy town, you'll find the finest in French champagne for a few bucks per bottle, Nike and Adidas sneakers at giveaway prices, bootlegged CDs of your favorite artists for two dollars, and even genuine Steinway pianos for the price of cut-rate concertinas.[1] Here you can also purchase land mines, antiaircraft guns, drugs, and rare and endangered animals. Ciudad Del Este is also a great place to pick up smuggled electronic goods from Miami or a stolen sports car from Brazil.[2] A Mercedes E55 AMG with a sticker price of $81,000 can be purchased here for less than $10,000. It is small wonder, therefore, that half of the 450,000 vehicles in Paraguay have been acquired illegally.[3] And here you are very likely (if you are a visiting gringo) to get killed.[4] There are from three hundred to four hundred reported murders a year. Most of the victims are listed as "tourists."[5]

Despite its population of 240,000, Ciudad Del Este contains fifty-five different banks and foreign exchange companies. US officials estimate that over $6 billion a year in illegal funds are laundered here—an amount equal to 50 percent of the gross domestic product of Paraguay.[6]

Another center for illegal activity, directly across the Parana River, is Foz do Iguaçu, Brazil, a city of 190,000. Here the cries to prayer from dozens of minarets are accompanied by the shrieks of the spider monkeys, the screams of the brown and red howlers, the cries of manakins and macaws, and the sound of the trumpeters. The golden domes and high minarets that rise from the mosques within Foz do Iguaçu appear as bizarre and incongruous in this lush jungle setting as the sight of the magnificent Teatro Amazonas Opera House within the scruffy port city of Manaus. The Muslims began to migrate here in 1975, at the outbreak of the civil war in Lebanon. By 2005 the Islamic population in the region swelled to sixty thousand.[7]

On any given day, the rusty iron Bridge of Friendship (*Puente de la Amistad*) that connects the two cities is packed with peddlers, merchants, and smugglers transporting boxes of "hot" goods from one country to the other without the slightest scrutiny from government officials or customs agents.[8] The goods include computers, plasma televisions, American cigarettes, surround-sound systems, digital cameras, laser tools, American sneakers, CDs, DVDs, designer jeans, and leather jackets. There are other goods as well, including tons of choice Colombian marijuana, hashish, and cocaine for drug dealers to transship to Puerto Paranagua on Brazil's Atlantic coast.[9]

Through the hordes of pedestrians, trucks, and vans snail their way across the four hundred-meter Bridge of Friendship. Upon arriving at the other side of the river, drivers change the license plates on their vehicles to match the country they have entered.[10]

With drugs passing through the crowded streets by the ton and assault rifles available for as little as $125, you are no safer here than in Ciudad Del Este. The murder rate in Foz do Iguaçu averages 250 a year, while hundreds of mysterious disappearances remain unrecorded.[11]

To reach Puerto Iguaçu, Argentina—the third city within the Triangle—you must cross the Tancredo Neves International Bridge from Foz do Iguaçu. Until 9/11, this small city with a population of twenty-eight thousand offered fugitives from justice and international terrorists a hot item that proved to be far more valuable than a Mercedes Benz, a Steinway grand piano, or even one kilo of choice Bolivian-produced cocaine. Here you could obtain counterfeit passports from Argentine officials for the measly sum of $5,000.[12] Such passports, under the visa waiver program, opened the portals to the Magic Kingdom, north of the Mexican border. The program made a valid Argentine passport the only document required for travel throughout the United States. In keeping with this goodwill agreement, tens of thousands of Argentines entered the United States and never returned to their native land.[13]

In 1983, Hezbollah (the "Party of God") became the first Islamic terrorist group to establish a base in Latin America. Funded, armed, and trained by the Iranian Revolutionary Guards, this group of radical Shiite Muslims came to realize that the lawless city of Foz do Iguaçu with its surrounding jungle would serve as an ideal place to train new recruits and to raise money for the *jihad* against the invading Israeli and American forces in Lebanon.[14] Not only was the city remote from scrutiny by US and Israeli intelligence agencies but also the Brazilian government recognized Hezbollah as a legitimate political party, and, therefore, all contributions to the organization were in full accordance with the law.[15]

Upon their arrival in the New World, the Hezbollah officials wasted little time in establishing business ties with the drug cartels from Colombia, Paraguay, Uruguay, and Ecuador, along with Latin American paramilitary groups such as the Revolutionary Armed Forces of Colombia (FARC) and Peru's Sendero Luminosos ("Shining Star").[16] The dealers and the revolutionaries needed guns that Hezbollah could provide through its connections with the Chechen Mafia, and Hezbollah sought cocaine that it could sell throughout the Middle East and Europe with the help of the Sicilian Mafia.[17]

As money, munitions, and recruits flowed into its South American cell, Hezbollah began to launch major assaults in Buenos Aires—including one against the Israeli Embassy on March 17, 1992, that killed 29 people and wounded 242, and another against the AMIA Jewish Center on July 18, 1994, that killed 86 and wounded 250.[18]

The attacks were purportedly orchestrated by Imad Fayez Mugniyah, who has been alternately described as the founder and the senior intelligence officer of Hezbollah.[19] Other than the place and date of Mugniyah's birth (Tayr Dibbuh, near Tyre, on July 12, 1962), little is known about him, save that his father, Sheikh Muhammad Jawah Mugniyah, was recognized in Lebanon as a Shiite scholar. Only two photographs of Mugniyah are known to exist, and both have been called into question. What's more, the elusive Shiite leader is said to have undergone a series of plastic surgical procedures to change his appearance.[20] US intelligence sources believe that he is of medium height (approximately five feet seven) and weighs between 145 and 150 pounds. Such sources, however, are even uncertain about the color of Mugniyah's eyes. According to Robert Baer, an ex-CIA officer who spent many years tracking Mugniyah, even the basic details of his childhood and education remain unknown. "Mugniyah systematically had all traces of himself removed," Baer told the New Yorker. "He erased himself. He had his records removed from high school and his passport application was stolen. There are no civil records in Lebanon with his name in them."[21] Regarding this elusive leader of Hezbollah, Magnus Ranstorp, a professor at the University of St. Andrews Centre for the

Study of Terrorism and Political Violence, said: "Imad Mugniyah is the very opposite of bin Laden. He has skills, is more professional, and operates as a faceless terrorist. We don't know what he looks like or where he is."[22]

In 1985, Mugniyah set forth the political platform of Hezbollah as follows:

1. The solution to Lebanon's problems is the establishment of an Islamic regime as the only government that can secure justice and equality for all of Lebanon's citizens.

2. The Hezbollah organization views as an important goal the fight against Western Imperialism and its eradication from Lebanon. The group strives for complete American and French withdrawal from Lebanon, including all their institutions.

3. The conflict with Israel is viewed as a central concern. This is not limited to the IDF [Israeli Defense Force] presence in Lebanon. Rather, the complete destruction of the State of Israel and the establishment of Islamic rule over Jews is an expressed goal.[23]

The platform reflected the ideology of Ayatollah Khomeini, the revered Shiite leader of the Islamic Revolution in Iran, who set forth a vision of a pan-Islamic theocracy headed by Islamic clerics. Hezbollah was one of Khomeini's creations—a terrorist group that could serve as a vanguard for the opposition against the invading Israeli army in Lebanon, where, the Ayatollah believed, the second phase of his glorious revolution was destined to take place. From the time of its inception, Hezbollah was armed and trained by the Islamic Republic of Iran through its Revolutionary Guard. It also received enormous financial support from Iran to the tune of $100 million a year.[24] Such support persisted even after the 1997 election of President Muhammed Khatami, who campaigned on a platform of tolerance and social reform.[25]

On April 18, 1984, Mugniyah made his first appearance on the FBI's "Most Wanted List" by masterminding the bombing of the US embassy in Lebanon. The bombing was remarkable in its economy and efficiency. It was accomplished by a single operative, who rammed into the embassy in a pickup truck that had been packed with explosives. The attack killed sixty-three people, including seventeen Americans, and six of the CIA's best Middle East experts.[26] It represented the first appearance of Mugniyah's contribution to the art of modern warfare: the suicide bomber.

Six months later, on October 23, another suicide bomber in yet another truck loaded with explosives smashed into the US Marine barracks outside the Beirut International Airport. The bomb produced the largest nonnuclear explosion in history with a yield equivalent to twenty thousand pounds of TNT, killing 220 Marines and 21 other US servicemen. It represented the bloodiest day for the US Marine Corps since the invasion of Iwo Jima during World War II.[27] Several hours later, a third suicide bomber struck at the headquarters of the French paratroopers in Beirut, leaving fiifty-eight dead in the mound of rubble.

As a complement to the bombings, Mugniyah initiated a host of other innovative terrorist acts, including the kidnapping of eighteen US officials. Three of these officials were tortured and killed, including William Buckley, the head of the CIA office in Beirut.[28] The US government now placed a $2 million bounty on Mugniyah's head.

By 1985 Mugniyah could relish the incredible results of his tactics: the withdrawal of US forces from Beirut, the collapse of the Israeli economy, the resignation of Israeli prime minister Menachem Begin, and the retreat of the mighty Israeli army from southern Lebanon in 1984. As a Shiite warlord, the elusive leader of Hezbollah had proven himself to be a worthy successor to Hassan-i-Sabah, the fabled "Old Man of the Mountains," who organized the "assassins" (the "drinkers of hashish") during the twelfth and thirteenth centuries. He had accomplished more than Muslim rulers such as President Gamal Abdel Nasser of Egypt, King Hussein of Jordan, and President Hafez Assad of Syria. Mugniyah and his small Party of God had defeated the mightiest military force that the Middle East had ever seen.

In 1996 the leader of Hezbollah performed an even greater feat: a union between Shiite and Sunni terrorists that would make possible the catastrophic events of September 11, 2001.

The bifurcation between Sunnis, who constitute 80 percent of the world's Muslims, and Shiites, who make up 15 percent of Muslims,[29] dates back to the assassination of Ali, the prophet Muhammad's son-in-law in 661. For thirteen centuries, the two groups remained actively hostile to one another. Wars between the two groups took place, most notably in the struggle between the Abbasids (the Sunnis) and the Fatimids (the Shiites) for control of Arabia, Egypt, and Syria. The bloody conflict persisted well into the twentieth century. The situation remained so volatile that bloody outbreaks between Sunnis and Shiites even erupted within the holy city of Mecca. A riot that occurred on July 31, 1987, during the annual *hajj*, left 480 pilgrims dead.[30]

The movement toward rapprochement among warring Muslims began to take place in the aftermath of an event, known among Islamists, as "al-Azma" or "the crisis." It refers to the first US-led invasion of Iraq in 1991—the so-called Gulf War under the administration of President George H. W. Bush. This event is superceded in Islamic infamy only by "al-Naqba" or "the calamity," that is, the establishment of the state of Israel in 1949.

In 1992, Hassan Abdallah al-Turabi, the spiritual leader of Sudan, set forth a theological compromise between the teachings of Sunni Egyptian Sayyid Muhammad Qutb (1906–1966) and the teachings of Ayatollah Khomeini (1902–1989). His proposal brought about a series of informal gatherings between Shiite and Sunni leaders in Khartoum, where spokesmen from both groups began to express ways in which they might work together to offset the Western menace.[31]

Eventually, the laying aside of differences between Sunni and Shiite terrorist groups came about, not as a result of a charismatic caliph or prominent imam, but rather by the action of Israeli prime minister Yitzhak Rabin, who is also known for his work on behalf of peace. In December 1992 Rabin ordered the deportation of 415 radical Palestinians to southern Lebanon. The exiled Palestinians were members of HAMAS, a Sunni terrorist group.[32]

HAMAS, an Arabic acronym for the Islamic Resistance Movement, was formed in 1988 to oppose all discussion of peace with Israel, including the Israeli-PLO negotiations.

In Lebanon, the Sunni terrorists of HAMAS were granted the shelter and protection of the Shiite terrorists of Hezbollah in accordance with the celebrated Muslim code of *milmastia* (hospitality). The exiled Sunnis responded to this gesture of goodwill by assisting the efforts of their Shiite hosts to gain a foothold within Israel[33]—something that Hezbollah had been unable to accomplish, since the Muslim population of Israel remained almost entirely Sunni and actively antagonistic to the presence of a Shiite party within the *waqf* ("the land of Palestine").

Other developments followed. Hezbollah began to train HAMAS in advanced bomb-making techniques along with the fine art of suicide bombing, a tactic that previously had been shunned by the Sunnis because of the Koran's injunctions against suicide.[34] The first HAMAS suicide bombing took place within a bus station in Hadera on April 13, 1994. The attack left five people dead and a score wounded.[35] An onslaught of other suicide bombings followed at sites throughout Israel in such rapid succession that it became difficult to discern if the attacks were being caused by HAMAS or Hezbollah.

This cooperation between Shiite and Sunni terrorists in Lebanon and Israel provided vivid proof that Muslims could set aside their differences and unite in a struggle against a common enemy. Throughout 1994, meetings between leaders of HAMAS and Hezbollah began to take place in Syria, Iraq, Iran, and other places throughout the Middle East. The stage was now set for a meeting between the two masters of Islamic terrorism: the Shiite Imad Mugniyah and the Sunni Osama bin Laden. According to US intelligence reports, agents of bin Laden made a series of calls from his Compact-M satellite phone to the Iranian Ministry of Intelligence and Security (MOIS) in 1995, offering to join forces in a *jihad* against

Israel and the United States.[36] Records of the calls were later unearthed by US intelligence officers during the investigation of the 1998 bombings of the US embassies in Kenya and Tanzania.

The monumental meeting between Mugniyah and bin Laden took place in 1995 at the headquarters of Ali Numeini, a Sudanese sheikh, in Khartoum. It proved to be an event that gave credence to the age-old Muslim proverb—"The enemy of my enemy is my friend." In October 2000, Ali Abdelsoud Mohamed, a former US Green Beret sergeant who pleaded guilty to conspiring with bin Laden in the bombings of the US embassies in Kenya and Tanzania, testified that he had provided security for bin Laden at the meeting.[37] Other face-to-face get-togethers between the two terrorist leaders followed, leading to a financial agreement regarding the flow of drugs from Afghanistan to Turkey through the northern region of Iran.

From June 21 to 23, 1996, Iran's Supreme Council for Intelligence Affairs sponsored a terror summit in Tehran in order to transform Hezbollah into "Hezbollah International," the new vanguard of the Islamic revolution that would coordinate the actions of Sunni and Shiite terrorist groups. The gathering attracted senior terror commanders, including Ramadan Shallah (the Palestinian Islamic Jihad), Ahmad Salah (Egyptian Islamic Jihad), Imad al-Alami and Mustafa al-Liddawi (HAMAS), Ahmad Jibril (Popular Front for the Liberation of Palestine), Abdallah Ocalan (the Kurdish People Party), Muhammad Ali Ahmad (al Qaeda), and Imad Mugniyah.[38] The summit resulted in the establishment of the Committee of Three that would meet on a monthly basis for the "coordination, planning, and execution of attacks" against the United States and Israel. The committee members were Imad Mugniyah, Ahmad Salah, and Osama bin Laden. The first meeting of the Committee rubber-stamped three terrorist operations that were already in the works: the bombing of the US barracks in al-Khobar, Dhahran (for bin Laden); the stabbing of a US female diplomat (for Salah); and the downing of TWA Flight 800 (for Mugniyah).[39] The new spirit of cooperation between the terrorist groups would result in bin Laden's "Declaration of War against the Americans Occupying the Land of the Two Holy Places" on August 23, 1996.

Developments occurred in rapid succession. Several top al Qaeda officials were invited to visit Hezbollah training camps in southern Lebanon. They returned to bin Laden's lair in Sudan with instructional videotapes for the blowing up of large buildings in major metropolitan areas. Ayman al-Zawahiri, bin Laden's top deputy and mentor, now became a frequent guest of Ali Fallahian, Iran's minister of intelligence and security, and Ahmad Vahidi, Iran's head of foreign terrorist organizations.

The alliance with Mugniyah proved to be highly beneficial for bin Laden. His terrorist organization's first major operation—the 1993 World Trade Center bombing—was a failure because the bombs were neither properly built nor properly delivered. With the help of Hezbollah, bin Laden learned how to rectify these mistakes as evidenced by the bombings of the US embassies in Kenya and Tanzania on August 7, 1998—bombings that bore the distinct signature of Imad Mugniyah.[40]

The same signature could be discerned on the attack on USS *Cole* by al Qaeda on October 10, 2000. The blast was caused by a "cone-shaped charge" that contained "moldable high explosives such as SEMTEX H," shaped to create a high speed, high temperature explosion. The first stage of the blast forced all the air out of the ship with tremendous force, thus creating a vacuum. The air, rushing back into the ship to fill the vacuum, created another tremendous explosion that caused further damage. It represented a device that had been uniquely developed by Mugniyah for terrorist attacks in Lebanon, Israel, and South America.[41] The bomb had been made in Lebanon under the protection of the Syrian army, who helped transfer it to the al Qaeda terrorists in Yemen.[42]

Mugniyah further imparted to bin Laden key techniques of successful terrorist management in such areas as the recruitment of choice operatives, the establishment of sleeper cells, the proper use of suicide bombers, and the tactical means of mounting simultaneous attacks.[43] All of these techniques would come into play on 9/11.

The alliance also proved to be an enormous boon for Mugniyah and his Iranian sponsors. In addition to continued assistance from HAMAS in mounting attacks within Israel and revenue from al Qaeda's drug route from Afghanistan to Bulgaria, the alliance served as the catalyst by which Iran gained access to the centrifuge technology of the A. Q. Khan Research Laboratories in Pakistan for the development of nuclear power plants and nuclear weapons.

With the launching of Operation Enduring Freedom and the invasion of Afghanistan, bin Laden experienced additional rewards from his alliance with Mugniyah that would ensure the survival of al Qaeda and the expansion of the war on terror into Muslim countries throughout the world.

In the wake of the bombing of Tora Bora in December 2001, more than five hundred al Qaeda and Taliban operatives scaled the mountains in the south along the Afghanistan-Pakistan border. They then cut through Afghanistan's southernmost provinces to head west toward Iran, where they found safe haven. The group included Saad bin Laden, Osama's eldest son; Yaaz bin Sifat, a top-ranking al Qaeda planner; Mohammed Islam Haani, the mayor of Kabul during the Taliban regime; Saif al-Adel, the military commander of al Qaeda; and Abu Musab al-Zarqawi, the al Qaeda commander who had been in charge of attacks in Europe.[44] Al-Zawahiri also traveled to Iran, where he was spotted in the disguise of an Iranian cleric with a black turban and a dyed red beard.[45] Upon their arrival, Saad bin Laden and al-Zawahiri were received by Osama's old friend Imad Mugniyah, who remained in Iran under the protection of the mullahs and religious leaders.[46]

The al Qaeda guests were placed in safe houses controlled by SAVAMA, the Iranian intelligence services. These villas, in southern Iran, with saunas and swimming pools, were lavish even by American standards.[47] Within Iran, they were able not only to move about freely but also to orchestrate a series of bombings in Saudi Arabia and continuous attacks on the coalition forces in Iraq. Thus, a monumental event had taken place that remained ignored by Western observers. The Muslim world of Sunnis and Shiites, thanks largely to Imad Mugniyah, had become united. "We have been screaming at them for more than a year now, and more since September 11th, that these guys all work together," an overseas operative told the *Washington Post*. "What we hear back is that it can't be because al Qaeda doesn't work that way. That is bullshit. Here, on the ground, these guys all work together as long as they are Muslims. There is no other division that matters."[48]

In 1995, as a result of the coming together of Sunni and Shiite terrorists, al Qaeda arrived on the American continent.

According to Brazilian intelligence sources, Osama bin Laden became the first member of his terrorist organization to venture forth to the New World. The al Qaeda leader reportedly spent three days in the infamous Triangle, where he visited the mosques of Foz do Iguaçu, met with the local leaders of Hezbollah, and attended meetings with leaders of the various drug cartels in Ciudad Del Este. News of the visit was first reported by CNN.[49] But its significance still eluded the press.

In 2003, eight years after the alleged visit, *Vega*, the leading Brazilian weekly newspaper, published photographs from a twenty-eight-minute video that had been taken of bin Laden to commemorate his visit. In the photographs, bin Laden appears in his traditional *shalwat kameez* (the long, loose-fitting shirt and baggy pants of the Afghan-Arab *mujahadeen*) with a well-trimmed goatee rather than the full beard he came to sport upon his return to Afghanistan in 1996.[50]

Imams and Arab community leaders within the Triangle profess to have no memory of the visit. "We are a peaceful working community," Sheikh Assayed Charis Sayed, the imam of the Prophet Muhammad Mosque in Foz do Iguaçu, said. "Bin Laden is a man of violence who's been asking for it."[51] Mijail Meskin, the Syrian consul within Ciudad Del Este, added: "The only thing it [the report of Osama's visit] has done is to increase uncertainty."[52]

Despite such statements, accounts of bin Laden's visit to the Triangle became so persistent that the municipal leaders of Foz do

Iguaçu, in a bizarre effort to attract tourists, began to run full-page ads in international newspapers with a photograph of bin Laden and the following caption: "When he's not blowing up the world, Osama bin Laden enjoys himself. If bin Laden risked his neck to visit Foz do Iguaçu, it means it is worthwhile! Everyone wants to see it. Why haven't you come?"[53]

Of Khalid Shaikh Mohammed's visit to South America, there remains no doubt. Proof came with the confiscation of his passport at the time of his arrest in Karachi on March 2, 2003. The passport bears a Brazilian tourist visa and shows that Khalid Mohammed, the al Qaeda military operations chief, entered Brazil from Pakistan on December 4, 1995, and departed for the Netherlands on Christmas Eve.[54]

During his three-week stay, Khalid Mohammed met with the leaders of Hezbollah in Ciudad Del Este and Foz do Iguaçu to make plans for attacks on the United States and Canada. One plan called for a series of terrorist attacks on synagogues, Jewish community centers, and landmark buildings in Ottawa in order "to undermine the Middle East peace process."[55] A second plan called for the blowing up of the Los Angeles International Airport. Neither plan came to fruition, thanks largely to the arrest of Ahmed Ressem of Montreal, who was caught trying to sneak massive amounts of explosives from British Columbia into the United States.[56]

Although the planned attacks never took place, Khalid Mohammed's visit proved to be a resounding success. By 1996 two Sunni terrorist cells appeared within the remote Brazilian city of Foz do Iguaçu. The first was a branch of al-Gama'a al-Islamiyya, a sister agency of al Qaeda, under the command of al-Saiid Ali Hassan Mokhles, who had been trained by al Qaeda for fifteen months in Afghanistan. From 1995 to 2002, Mokhles kept in close contact with Mohammed and other members of al Qaeda's *Shura* (consultation council). Copies of the communiqués between the two terrorists were discovered by US intelligence officials within Khalid Mohammed's headquarters in Karachi.

The second cell represented a branch of Al-Gama'a al-Islamiyya, another of al Qaeda's hydra heads. It was placed under the command of Mohamed Ali Aboul-Ezz al-Mahdi Ibrahim Soliman, an Egyptian national and close friend of Ayman al-Zawahiri.

The visit by bin Laden's military operations chief and the continued communication between al Qaeda officials and radical Muslims within the Triangle eventually led to the creation of a seamless terrorist highway that ran from the Port of Iquique in Chile; through the tri-border area of Brazil, Argentina, and Paraguay; north through the Brazilian jungle; then branching out to Surinam, Guyana, Venezuela, Colombia, Panama, Honduras, and Mexico. From Mexico the route led to terrorist alley and other points of entry into the United States that remain under the control of the Mara Salvatrucha gang.

Bin Laden had discovered within the remote tri-border region of Argentina, Paraguay, and Brazil a setting for his terrorist activities that was far better than the mountainous terrains of Afghanistan or the remote desert regions of Sudan, where his every move had come under scrutiny of US, British, Israeli, Saudi, and Russian intelligence. It was a country beneath the radar—a country largely ignored by the international intelligence community—a third world country that few believed could give rise to cataclysmic events. The Triangle offered unguarded borders, numerous waterways, and more than one hundred hidden airstrips. It was a place where millions could be raised and laundered for the *jihad* and where thousands of recruits could be trained in guerrilla warfare.

What's more, within the dense jungles of the region, al Qaeda could establish laboratories for producing and testing weapons of mass destruction.[57] Brazil possessed the world's fourth-largest reserve of uranium, a large nuclear facility that could be used for the enrichment of uranium in Resende, and two nuclear power plants (Angna I and II).[58] For this reason, the Latin American country represented a perfect place to obtain low-grade uranium for the creation of dirty bombs and, eventually, weapons-grade or highly enriched uranium for use in tactical weapons.

Such prospects did not represent idle pipe dreams. In the past,

Brazil had engaged in nuclear proliferation for profit. From 1979 to 1990 the Brazilian government supplied Saddam Hussein with hundreds of tons of natural and low-enriched uranium, reactor technologies, equipment, and training for his nuclear program in Iraq.[59] By 2001 the Triangle became a center not only for such illicit activities as arms dealing, drug trafficking, and money laundering but also (thanks to al Qaeda) the unique crime of uranium smuggling. Proof of this came in August 2004 when police seized a load of 1,320 pounds of uranium and thorium ore in a pickup truck seventy-five miles from Mocopa, near the mouth of the Amazon.[60]

Other al Qaeda luminaries made their way to the Triangle, including Ramzi bin al-Shibh, a key planner of 9/11. While visiting South America, al-Shibh reportedly met with Imad Mugniyah at the Maskoud Plaza Hotel in Sao Paulo, Brazil.[61]

Khalil bin Laden, Osama's younger brother who owns a twenty-acre estate in Winter Garden, Florida, also made frequent trips to the area, including outings to Minas Gerais, the site of an al Qaeda training camp within the outback of Brazil, where he engaged in "suspicious business activities."[62] Khalil developed close relationships with Brazilian officials, married a Brazilian woman, and became Brazil's honorary consul in the Saudi Arabian city of Jeddah in 1998.[63]

On September 20, 2001, Khalil bin Laden, although named as a key terror suspect by South American officials, was permitted to board a chartered jet from Saudi Arabia, along with fourteen other members of the bin Laden family, and depart from the United States. The young bin Laden had never been questioned, let alone detained, by US intelligence officials.[64] On March 6, 2003, Bosnian police seized documents from the Muslim Benevolence International Foundation (BIF), a terrorist front, in Sarajevo. One document contained a list of the top twenty financial supporters of al Qaeda. Khalil bin Laden's name appeared near the top of the list.[65]

As an honorary consul, Khalil bin Laden may have served in

establishing a connection between the Brazilian governmental officials and Dr. Abdul Qadeer Khan.

At the close of 2002 came allegations that Dr. Khan had sold his centrifuge technology to the Brazilian government of President Luiz Inacio Lula da Silva, affectionately known by his supporters as "Lula." A lifelong Marxist and ally of Fidel Castro of Cuba and Hugo Chavez of Venezuela, Lula is a cofounder the Forum of Sao Paulo, a group that supports anti-American activities throughout the world.[66] He is also one of Latin America's sharpest critics of Operation Enduring Freedom (the US-led invasion of Afghanistan) and the war in Iraq.[67]

While campaigning for Brazil's highest office, Lula criticized his country's compliance with the 1970 Nuclear Nonproliferation Treaty by saying: "I imagine this would make sense only if all countries that already have nuclear weapons also gave them up. It is not fair that developed countries, which have nuclear weapon technology, demand that others not have them or deactivate what they have. All of us developing countries are left holding a slingshot while they have atomic bombs."[68]

Upon his election in 2002, Lula launched his program to transform Brazil into a technological and military power. By 2004 the Resende nuclear facility announced its capacity to enrich uranium by means of ultracentrifugation.

The news came from left field (i.e., Latin America) and sent shock waves through the Bush administration that had remained fixated on nuclear developments in Iran and North Korea. Then secretary of state Colin Powell visited Brazil and expressed his confidence that Brazil possessed no plans to develop nuclear weapons.[69]

Powell's statement did little to allay fears that Brazil had purchased its technology from Dr. Khan and that the Latin America country was in the process of highly enriching uranium to produce an arsenal of nuclear warheads. As noted, Science magazine predicted that Brazil by 2006 would be capable of producing five to six warheads a year.[70] Such suspicions were fueled by the fact that Brazil refused to grant IAEA inspectors the right to conduct a full inspection of the Resende facility.[71] The inspectors were merely allowed to view parts of the centrifuges while the Resende officials kept other parts under cover.[72] "They [the IAEP inspectors] are specifically worried about the Khan network being one of the sources of the program," said Henry Sokolski, former Pentagon official and head of the Nonproliferation Policy Education Center in Washington. "I can't tell you how I know, but I know."[73]

The secrecy surrounding the Resende facility caused Constantine Menges, a senior fellow of the Hudson Institute, a think tank in Washington, DC, and a former national security advisor to President Reagan, to express fears that a nuclear-armed axis of evil between Brazil, Venezuela, and Cuba was taking shape within the Americas. "Lula's a supporter of terrorism," Menges said. "He will, I believe, permit covert support to be given to bring about anti-American regimes in Bolivia, Colombia, Ecuador, and Peru.[74]

CHAPTER NINE
Too Little, Too Late

In key countries, percentages of citizens with a negative view of the United States are rising: 68 percent in Mexico, 62 percent in Argentina, 42 percent in Brazil, and 37 percent in Chile.
—Latin American Security Challenges,
US Naval War College, Newport, Rhode Island, 2004

I certainly wouldn't put it past al Qaeda to use Latin America as a route to place its assets in the United States.
—Former CIA director James Woolsey, November 2, 2001

Fidel for me is father, a companion, and a master of the perfect strategy.
—Hugo Chavez, president of Venezuela, October 20, 2004

Iran and Cuba, in cooperation with each other, can bring America to its knees. The US regime is very weak, and we are witnessing this weakness from close up.
—Fidel Castro, president of Cuba, speaking to
Muslim students at the University of Tehran, May 1, 2001

Bin Laden opened the door to his al Qaeda haven in the New World not only to friends and family from the Middle East but also to his business associates and radical Muslim comrades from the former Soviet Union: the Chechen Mafia. The relationship between bin Laden and the Chechen Mafia that had been forged during the Azerbaijan-Armenia conflict and the establishment of the Abkhaz drug route was now bearing fruit. By 1996 the Chechen Mafia, from the frozen steppes of Russia, had made themselves right at home in the tropical climate of Ciudad del Este and Foz do Iguaçu with members of other international crime families—the Paraguayan cartel of Pedro Juan Caballero, the Japanese-Brazilian cartel, the Turkish cartel, and the Chinese Mafia (14-K Triad, Pak Lung Fu)—and began engaging in lucrative arms-for-drugs deals.[1] Overnight, Brazil, a nonproducing country for illegal drugs, became the leading transit center for the cocaine trade from Bolivia and Colombia headed for the United States and Europe, via Cape Verde, the Ivory Coast, and South Africa.[2] Most of this trade came under the control of the Chechens, who managed to terrorize not only the warlords and drug lords of Colombia and Bolivia but even their fellow Islamic terrorists within the Triangle.

In 1997 the CIA asked its counterparts in SIDE, Argentina's Secretaria de Inteligencia del Estado (Secretariat for State Intelligence) to infiltrate the Triangle in order to obtain information on the workings of the Islamic extremist groups. During the course of the operation (which was known by the codename "Centauro"), SIDE conducted a comprehensive surveillance of the various terrorist cells by means of phone taps, mail interception, covert filming, and interrogation of hundreds of suspects. The evidence showed that al Qaeda had established virulent cells in the area. These cells served to raise millions for the *jihad*, to manufacture counterfeit documents, and to operate a large training camp for new recruits in the Brazilian province of Minas Gerais, not far from the tri-border area.[3] Further surveillance uncovered the fact that al Qaeda was working not only with fellow Sunni organizations but also with Shiite groups, such as Hezbollah, for a massive attack on the United States.[4]

The SIDE officials came to focus their attention on al-Saiid Ali Hassan Mokhles, the Egyptian national, who served as the leader of the al-Gama'a al-Islamiyya cell in Foz do Iguaçu. They discovered that Mokhles, who had been trained in Afghanistan and sent to Brazil by bin Laden, had been engaged with members of his cell in raising funds for the *jihad,* planning terrorist attacks (including an attack in Luxor, Egypt, that resulted in the massacre of sixty-two tourists), and counterfeiting passports and other official documents for use by al Qaeda sleeper agents.[5] They came to realize that Mokhles worked in close association with the local leaders of Hezbollah and HAMAS so that there remained "no operational differences" between the Sunni and Shiite terrorists within the Triangle.[6] SIDE officials further uncovered the fact that Mokhles kept in constant contact with his al Qaeda superiors in Afghanistan, including al-Zawahiri and Khalid Shaikh Mohammed.[7]

On January 26, 1999, Argentine officials, working on a tip from SIDE, arrested Mokhles at a border control station in El Chuy, Uruguay, as he and his family attempted to leave Brazil with a few of his phony passports and flight tickets to London.[8]

Shortly after the arrest of Mokhles, the relationship between the CIA and SIDE suffered a complete breakdown when US intelligence officials accused SIDE of leaking a photo of Ross Newland, the CIA bureau chief in Buenos Aires, to a local newspaper.[9] Newland was recalled to the United States; the CIA's ties with SIDE were severed; and the evidence uncovered by Operation Centauro remained filed and forgotten.

Concerns about the area were not even raised among CIA officials the week before 9/11 when Abdel Fatta, a young Moroccan living in Foz do Iguaçu, handed a letter to his lawyer with instructions that it be delivered to the US embassy. The letter contained a warning about a series of attacks that were to take place in Washington, DC, and New York on the eleventh of September.[10] Fatta, as it turned out, was an al Qaeda member who experienced a change of heart. He remains, at this writing, safely sequestered within a prison cell in Brazil.

In the wake of 9/11, the Triangle came to resemble Casablanca at the outbreak of World War II—a place where local intelligence and police agencies were joined by operatives from SIDE, the CIA, the FBI, and even Israel's Mossad. Information became peddled on every street corner. Arrests were made. A Paraguayan SWAT team raided the office of Assad Ahmad Barakat, the owner of the largest shopping mall in Ciudad Del Este, and seized financial statements showing the transfer of more than $50 million to Hezbollah in Lebanon and Syria. Some of these transfers were made from the port city of Iquique, Chile, to a Chase Manhattan Bank account under the name of "Mr. Barakat" in Ciudad Del Este. The police also found within Barakat's office several boxes of al Qaeda promotional videos and training tapes for suicide bombers.[11]

Further raids led to further arrests of Hezbollah militants, including that of Ali Khalil Mehri, who operated a software-pirating scheme in the area, and Sobhi Fayad, who purportedly extorted business owners within the Triangle to give 20 percent of their income to the terrorist group.[12]

Several Sunni terrorists with ties to bin Laden were arrested, including Mohamed Ali Aboul-Ezz al-Mahdi Ibrahim Soliman and Mohamed Enid Abdel Aal. Both were prominent members of al-Gama'a al-Islamiyya, one of al Qaeda's sister agencies.[13] Aal was collared after he fled from the Triangle and attempted to reach a FARC (Revolutionary Armed Forces of Colombia) guerrilla base in Colombia.[14] Soliman and Aal were charged with a myriad of criminal offenses, including money laundering, extortion, and drugs-for-arms dealing.

Ali Nizar Dahroug, another member of al-Gama'a al-Islamiyya within the Triangle, was arrested after US intelligence officials discovered his name within the address book of Abu Zubaydah, a high-ranking al Qaeda official who was captured in Afghanistan. Paraguayan officials discovered that Dahroug owned a small perfume shop in Ciudad Del Este with assets less than $2,000 and that he was wiring more than $80,000 a month to banks in the United States, the Middle East, and Europe.[15]

The raids and arrests produced enough evidence for Paraguayan interior minister Julio Cesar Fanego to announce that more than $500 million, enough to fully realize bin Laden's dream of an American Hiroshima, had been raised by the terrorists within the lawless Triangle of Argentina, Paraguay, and Brazil in less than two years (1999–2001).[16]

There were other discoveries. One involved proof that the Paraguayan consul stationed in Miami, Florida, had sold a batch of three hundred passports, visas, and shipping documents. This included 19 passports to terror suspects from Egypt, Syria, and Lebanon who planned to move to Ciudad Del Este.[17] A second discovery concerned a plot by a Hezbollah cell in Foz do Iguaçu to assassinate Mexican president Vincente Fox and members of the Mexican Senate.[18]

By and large, however, the renewed cooperation between SIDE and the US intelligence agencies proved to be too little, too late. The two malignant cells that had been planted by bin Laden within the belly of South America had metastasized so that the cancer of al Qaeda and radical Islam began to appear in various locations throughout South America.[19]

For starters, there were other locations within Brazil and Paraguay. As soon as the intelligence agencies began to comb the Triangle, swarms of Muslims with ties to radical Islam migrated to Pedro Juan Caballero on the northern border with Brazil, an area renowned for its potent marijuana and nonexistent policing.[20] By 2002 evidence began to surface that the Brazilian town of Chui, near the Uruguay border, with a population of fifteen hundred Muslims, had become a haven for al Qaeda and that the mayor purportedly served as an agent for bin Laden.[21]

Within Paraguay, officials of the Silvio Pettirosi International Airport in Asuncion began to report regular arrivals of Middle Easterners with European identification papers. When questioned by authorities, the newcomers spoke fluent Arabic but were unable to speak the language of their native land as listed on their passports.[22]

The arrest and interrogation of an Afghanistan-trained member of the Islamic Group of Egypt in Uruguay on February 27, 1999, served as the first warning that the country had been infested by bin Laden's terrorist group.[23] The virulence of the cell was verified by the US decision on April 6, 2001, to suspend its diplomatic missions in Uruguay, Ecuador, and Paraguay because of the likelihood of an al Qaeda attack.[24]

Then there was Chile. Muslim terrorists, largely from Hezbollah and HAMAS cells in Foz do Iguaçu and Ciudad Del Este, made their way to Iquique on Chile's northern Pacific coast, where they set up their money laundering operations. By 2002 the Bush administration designated the once sleepy resort town as a terrorist "hot spot" within South America, second only to the Triangle.[25]

Within a year, Peru loomed as a country of even greater concern, after investigators uncovered evidence that Vladmiro Montesinos, the former head of Peruvian intelligence, had allowed bin Laden to acquire a host of homes and apartments to serve as safe havens for members of his terrorist network. In one of the secretly recorded conversations that led to his arrest on charges of money laundering, bribery, and terrorism, Montesinos told the mayor of the Peruvian port city of Callao that "bin Laden's center of importance in Latin America is here in Lima."[26] The former spy chief explained to the mayor that bin Laden's center in Lima had been established "to act against Americans in Argentina, Brazil, Chile and the rest of South America."[27] By the close of 2001, Peruvian authorities came to realize that the corrupt Montesinos had made millions through a clandestine agreement with the al Qaeda chieftain.[28]

Before revelations about the Peruvian strongman came to light, Montesinos was considered a military intelligence asset by the US and received support from the Clinton administration. The truth of the jailed spy chief's words about the pivotal importance of Peru to the *jihad* became supported by the appearance of al Qaeda cells in the central Andes, the Ayacucha highlands, the Huallaga valley, and the northwest Amazon jungle. Many of these cells remain under the protection of the Shining Path (Sendero Luminosos), a paramilitary Maoist organization, which has increased in strength and number after 9/11.

Known members of al Qaeda
on the
FBI's Most Wanted Terrorists list

(More information on these individuals is available at
http://www.fbi.gov/mostwant/terrorists/fugitives.htm.)

Osama bin Laden

Ayman al-Zawahiri

Zubayr al-Rimi

Karim el-Mejjati

Jaber A. Elbaneh

Dr. Mohammed Khan

Adnan G. el-Shukrijumah

Abderraouf Jdey

Faker Ben Abdelazziz Boussora

Abdelkarim Hussein Mohamed al-Nasser

Abdullah Ahmed Abdullah

Muhsin Musa Matwalli Atwah

Ali Atwa

Anas al-Liby

Ahmed Khalfan Ghailani

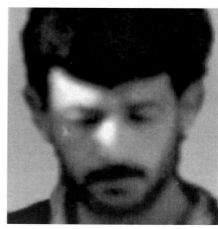

Hasan Izz-al-Din

Amer el-Maati

Azzam the American

Adam Yahiye Gadahn

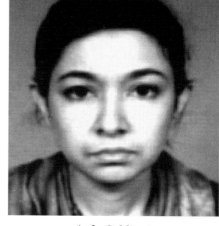

Aafia Siddiqui

The risk of Bolivia as a terrorist threat was raised from low to moderate in 2002 when the Bolivian government collared nine Bangladeshis in La Paz who were involved in a plan, uncovered by the French police, to hijack a jumbo jet for an attack on an American target in Argentina. The Bangladeshis, as it turned out, were members of a virulent al Qaeda cell with tentacles throughout Bolivia and ties to the country's lucrative illegal drug trade.[29]

Ecuador, rated Latin America's second-most-corrupt country (bested only by Paraguay) by the CIA, also became a haven for Hezbollah, HAMAS, and al Qaeda. Take the case of Lago Agrio, once a picturesque village within a tropical rain forest, now a fetid wasteland after years of oil drilling and exploration by the company that is now Chevron Texaco Corporation. From 1964 to 1992 Texaco reportedly dumped eighteen billion gallons of drilling by-products into open pits and waterways.[30] The toxic town, with its rotting infrastructures and lack of policing, could no longer attract oil speculators, international entrepreneurs, or wealthy American tourists. Texaco terrorism gave way to the terrorism of radical Islam, as Lago Agrio, with its population of twenty-eight thousand, became a lawless center for drug trafficking, arms dealing, smuggling, and money laundering—and with a murder rate in excess of one hundred a year, one of the most dangerous places on earth.[31]

And then there was Colombia, where, according to Gen. Rosso Jose Cerraro, Colombia's former chief of police, hundreds of Arab terrorists entered in recent years with false identification papers from other Latin American countries. They came to work with the Cali and the Eastern Plains Cartels in the planting and harvesting of poppy fields for the production of choice Number Four heroin—the drug of choice in Europe and the United States.[32] The poppy crops in the mountains of Colombia were planted in the same exact manner as the mountains of Afghanistan, giving proof positive to the symbiotic relationship that exists between organized crime and international terrorism.[33] Thanks to radical Islam, Colombia became the third-largest exporter of heroin in the world.

Within Colombia, the terrorists received support and protection from the Revolutionary Armed Forces of Colombia, known by its Spanish initials FARC. FARC is the military wing of the Colombian

Communist Party and represents the country's largest and most radical rebel group with a history of bombings, mass executions, and kidnappings. In 1998 FARC had won several concessions from the Colombian government of President Andres Pastrana, including the right to create a 16,200-square-mile "safe haven" zone (*Zona Despoja*) for its operations.[34]

In the aftermath of 9/11, Jorge Bricono (aka Mono Jojoy), the leader of FARC, called upon the seventeen thousand members of his group "to make them [the North Americans] feel the pain which they have inflicted on other peoples." Bricono urged FARC "to take away any means in order to defeat them [the citizens of the United States]. Reach out to North Americans who are unhappy and organize them. Reach out to black North Americans and make them see they are discriminated against."[35]

In 2001 a House subcommittee found that Colombia, now home to thirty-one groups blacklisted by the United States as terrorist agencies, represents "a breeding ground for international terrorism equaled only by Afghanistan."[36]

By 2002 US officials isolated cells of al Qaeda in Tobago, Trinidad, French Guiana, and Guyana (where Adnan el-Shukrijumah's father served as the imam of a radical mosque and where the elusive Adnan sought sanctity after being identified by the FBI as "the next Mohammad Atta").

But the extent of the spread of radical Islam throughout many northern regions of South America became best exemplified in the case of Surinam, a small country—the size of the state of Georgia—with a population of 437,000. By 2004, 35 percent of the population was Muslim. Along with the minarets and mosques appeared cells of radical Islam, including a Surinamese al Qaeda link to Jemaah Islamiyah, the terrorist organization responsible for the 2002 Bali bombing. The Chechen Mafia also set up operations in Surinam, where the price for an automatic rifle, such as an AK-47 from the Russian arsenal, became one kilogram of cocaine.[37]

But few places on earth proved to be more worrisome than Venezuela. On January 5, 2003, Venezuelan air force major Juan Diaz Castillo, who served as the personal pilot of Hugo Chavez, president of Venezuela, dropped a bombshell by stating that Chavez had awarded al Qaeda a gift of $1 million in the wake of 9/11.[38] The money, according to Castillo, was channeled through Walter Marquez, the Venezuelan ambassador to India.[39]

Few observers expressed shock at the news, and Castillo's testimony about the contribution failed to receive coverage by most major US news outlets. Such oversight among editors and journalists, if not justifiable, was at least understandable. Chavez's hatred of the United States was well known. On September 12, 2001, his supporters celebrated the terrorist attack by burning an American flag in Caracas's Plaza Bolivar.[40] When the US-led coalition forces invaded Afghanistan, Chavez decried them as "terrorists."[41] And in 2002, along with his friend and mentor Cuban president Fidel Castro, Chavez made state visits to such radical Islamic countries as Libya, Iran, and Iraq, where he signed cooperative agreements as Castro told thousands of cheering Muslim students: "Together we will bring America to its knees."[42]

Despite the lack of press coverage, few developments could have been more nightmarish for the United States than the union of Chavez and Castro. By the close of 2002, the Bush administration learned that Cuba's Intelligence Directorate (known by its Spanish initials DGI) had taken over many of the functions of Venezuela's Directorate for Intelligence, Security, and Prevention (DISIP) and that Ramon Rodriguez Chacin, Venezuela's interior minister, had covered up the identities of terrorists—many from the Middle East—who had passed through the country in order to deceive US terrorism investigators. Other developments came to light. US investigators discovered that thousands of such foreigners, including "suspicious individuals" from the Middle East, had been granted Venezuelan identity papers (cedulas and passports) by the Chavez administration to assist in their migration to Central America, Mexico, and the United States.[43] "I quit my job when I got tired of doing dirty work for Chavez with the Cubans looking over my shoulder," said Gen. Marcos Ferreira, who had served as director of Venezuela's Border Control Service.[44]

From General Ferreira, Major Castillo, Brig. Gen. Nestor Gonzalez Gonzalez, and other high-ranking Venezuelan military defectors, Bush administration officials reportedly learned that Cuba had established a biotechnological laboratory in San Antonio Los Altos near Caracas for the development of biological weapons against the United States.[45] These reports were verified by Carl Ford Jr., assistant secretary for research and development at the State Department, who told the Senate Foreign Relations Committee on June 5, 2002: "Cuba has provided dual-use technology to rogue states. Such technology could support bioweapons programs in those states. We feel very confident about saying that they're working and have been working on an effort that would give them a BW—limited BW offensive capacity. And that's serious enough to tell you."[46]

The bad news concerning Venezuela kept getting worse. On February 13, 2004, a Muslim extremist with suspected ties to al Qaeda was arrested at London's Gatwick Airport after a grenade was found in his luggage. His ticket showed that he flew to London from Colombia. But British authorities later learned that he was a Venezuelan who had been trained at an al Qaeda cell on Venezuela's Margarita Island, a place that once served as a haven for American tourists.[47] Bin Laden's group was alive and well and less than a three-hour flight from the United States.

A covert operation was unleashed by British intelligence. MI6 agents were dispatched to Venezuela to prevent the al Qaeda cell from running guns and cocaine into the United Kingdom. The guns originally came from Cuba's arsenal of Soviet weapons that had been provided to Castro during the cold war and given to the Marxist Revolutionary Armed Forces of Colombia (FARC). The agents discovered that other, updated weapons were arriving for FARC and the radical Muslims on fishing boats off the virtually unguarded coastline of Venezuela.[48] The weapons, including surface-to-air missiles, were coming from North Korea.

The finding sent shock waves through the US intelligence community. Little attention had been paid to Chavez's growing ties to China and North Korea—not even when Chavez proclaimed himself to be a "Maoist"[49] or when Hector Navarro, his minister of education, extended Venezuela's salute of "solidarity" to such "friendly nations" as "Algeria, Cuba, Iran, and North Korea."[50]

Now North Korea was providing Venezuela with high-tech weapons. The two countries had become allies not only in political ideology but also in acts of military aggression. This raised pressing questions. If weapons of conventional warfare were coming from North Korea, could other far-more-dangerous materials be coming from the communist country as well? Was Chavez hoping to benefit from North Korea's newly created nuclear technology? Could the nuclear proliferation of Dr. A. Q. Khan have extended to yet another rogue nation? No one possessed the answers. "It is too early to determine with certainty what the nuclear weapons plans of Hugo Chavez are," Brigadier General Gonzalez said. "But his reaffirmation of support for North Korea is a troubling sign. I personally know Chavez very well, and he is capable of anything."[51]

The developments within South America showed that radical Islam was pressing ever northward, closer and closer to the back door of the United States. In December 2003 Canada and Interpol informed Mexican officials that al Qaeda terrorists had set up cells within their country to mount their next attack on US soil. "The alert has been sounded," Jose Luis Santiago Vasconcelos, Mexico's top anticrime prosecutor, told the Associated Press.[52] An investigation was launched, resulting in the cancellation of three Aeromexico flights from Mexico City to Los Angeles.

In May 2004 Honduran security minister Oscar Alvarez admitted that police officials had uncovered evidence that al Qaeda terrorists had infested his country and were recruiting Hondurans for the next attack on American soil.[53] Although he did not mention details, Alvarez said: "This is very serious. We are talking of Honduran citizens being prepared to commit terrorist acts. We understand that the recruits undergo a kind of 'brainwashing' in order to become Muslim followers and possible martyrs in attacks against civilians. The investigations are intended to keep track of people who have been granted a scholarship to ascertain whether their behavior undergoes changes and what kind of communications they continue to have with groups which the government has declared to be terrorist."[54]

In July 2004, two months after Security Minister Alvarez issued this warning, Adnan el-Shukrijumah, who has been commissioned to commandeer the next 9/11, was spotted at an Internet café in Tegucigalpa, the capital of Honduras, with members of Latin America's most violent street gang.[55]

Concerns about the situation south of the border were further increased with the arrest of Farida Goolam Mohamed Ahmed, a forty-eight-year-old South African woman, who was collared while trying to board a flight at the McAllen Miller International Airport in Texas. The woman was flagged by a routine computer database inquiry initiated by a Customs and Border Protection officer. Although several pages within her passport were missing, authorities were able to discover that Ahmed had traveled from Johannesburg on July 8, via Dubai, United Arab Emirates, to London, then to Mexico City on July 14. The countries through which she traveled do not require South African citizens to have tourist visas.[56] From Mexico, she waded across the Rio Grande with hundreds of other illegal immigrants. Mexican officials said that Ahmed was not stopped upon entering Mexico because her name did not appear on any watch-list for international terrorists. The arrest was aired in the national press and quickly forgotten as a matter of small consequence.

But the case of Farida Goolam Mohamed Ahmed represented a matter that should have raised screams of alarm within the halls of Congress about the state of national security and the problem at the two thousand-mile border between Mexico and the United States.

Ahmed was a terrorist courier who provided information and instruction from the *mujahadeen* in the Middle East to an al Qaeda cell in New York. Several of her communiqués concerned plans for a major bombing operation in midtown Manhattan. What's more, she had crossed the border into the US not just once or twice but on hundreds of other occasions over an undisclosed period of time.[57] Yet Ahmed the faithful mule had never been detained by the US border patrol or questioned by security officials at the Texas airport.

Fortunately Ahmed was also a talking mule, and her interrogation led to the arrest of several al Qaeda agents by Mexican authorities.[58] These agents, along with Ahmed, were transferred to one of the CIA's clandestine terrorist detention facilities, where they remain at this writing.

CHAPTER TEN

The Terrorists and the Gangbangers

Americans would be shocked to learn that state and local officers are routinely releasing illegal aliens during the course of their normal duties. In fact, many illegal aliens are not deported even after serving prison sentences for criminal convictions because federal and state and local officials are not talking to each other.

—Senator Jeff Sessions (R-AL), press release,
November 20, 2003

We received intel that seven males of Middle Eastern descent had entered the United States illegally, and they were here to carry out a terrorist attack on this country. I'd never seen the agency go to the level that we did as far as treating it with such urgency. When we got the information, we immediately sent agents to patrol the highways. We had never done anything like that before. And like I said before, within a week, watching the nightly news and the terror threat was raised, and they actually showed pictures of these seven men who had reportedly entered the country. And that's when it kind of really hit me for the first time that it's really happening.

—Border Patrol agent, speaking anonymously to KVOA-TV
reporter in Tucson, Arizona, November 11, 2004

While the vast majority of illegals from the Middle East are not terrorists, the fact that tens of thousands of people from that region—and millions more from the rest of the world—can settle in the U.S. illegally means that terrorists, who wish to do so, face few obstacles. We cannot protect ourselves from terrorism without dealing with illegal immigration.

—Stephen A. Camarota, director of research,
the Center for Immigration Studies, November 2004

The Salvatrucha gangs are very serious, very vicious, and we have confirmed that they have had contact with al Qaeda.

—Congressman Solomon Ortiz (D-TX), September 2, 2004

Felix Ortega, aka "Tres Ojos" ("Three Eyes"), is "yoked," "ripped," and "jacked" from weight lifting after an eighteen-month stint in a slammer for aggravated assault. Even slouched in a chair at a table of the Uno restaurant within the District of Columbia's Union Station, he radiates menace and raw physicality. The long, thick veins that run down the sides of his neck seem to have sprouted like coaxial cables (along with his square, flat face) from a massive clavicle. The dense and well-sculpted muscles in his shoulders and arms bulge like cantaloupes when he takes a sip of his Corona or a drag from his Lucky Strike.

Ortega's eyes are brown, unblinking, and feral. He glares at the "suits" who sit across from him at the table with the civility of a pit bull. A mustache, so incredibly thin it may have been painted by a Lilliputian, is scarcely visible above the subject's thick lips that appear to have been pressed into a sullen frown. The frown testifies to Oretga's "attitude"—a cherished quality of gangbangers (bangers is slang for gang members who engage in acts of violence).

Ortega is dressed and groomed for the "informal" interview as if he walked out of central casting for an R-rated update of *West Side Story*. He wears a "wife-beater" shirt and voluminous baggy jeans that bunch at his ankles above a pair of black Frankensteins (heavy leather gym shoes). Around his neck is a gold chain so thick and heavy that it might be used to tow a small Toyota truck out of a sink hole. He sports other jewelry, including a twenty-four-karat skull ring and dangling silver earrings in the shape of stilettos.

The gangbanger's head has been shaved, and his body has been covered with lurid and poorly drawn tattoos of naked women, writhing snakes, fire-breathing dragons, death heads, dice, and pentagrams. His face, too, has been adorned with purple and red tattoos. A spider's web stretches from cheekbone to cheekbone over the bridge of his nose, while the name "Soreno" (Southerner) has been etched on his forehead. The letter *M* appears on his right bicep, the letter *S* on his left; the number 1 on his right tricep, the number 3 on his left. MS-13 is the gang's codename

Ortega professes to hate all Americans, including "peckerwoods" (whites), "niggers," and "spics" (fellow Latinos who have abandoned the barrio for the suburbs). "The United States is a country of putos [pimps] and putas [whores]. You can say to an Americano that his mother is a puta and his father is a puto and he will just smile at you as though you are paying him a compliment. Gringos have no cajoles. Most of them are now becoming jotos [homosexuals]. They think this is politically correct. I think it might be evolution." The statement represents Oretga's attempt to display his erudition and wit. It causes him to flash a ghost of a smile. His two front teeth are missing.

Asked about the moniker "Tres Ojos," Ortega says: "My third eye is my pecker. It always is on the lookout for pussy." The smile now becomes a smirk.

"Tres Ojos" is engaged in dissin'—showing contempt toward gringos. This is a ritual that every *jefe* (gang leader) must perform in order to maintain self-respect.

Ortega is a member of Mara Salvatrucha. The gang was formed as a result of the civil war that raged in El Salvador during the 1980s. The war killed more than one hundred thousand Salvadorans and left millions destitute and homeless. Hundreds of thousands made their way to the United States and settled initially in the established Hispanic neighborhoods of the "Rampart" section of Los Angeles.[1]

The newcomers found themselves at first shunned and later victimized by the various Hispanic gangs—the Crips, the Bloods, the Latin Kings, the Mexican Mafia, and the Crazy Riders. As a result, they formed Mara Salvatrucha. In El Salvador, a sizable number were members of La Mara, a violent street gang, who named themselves after a particularly aggressive breed of African soldier ants.[2] Others belonged to the left-wing Farabundo Mara Liberation Front, a paramilitary group of peasants that had been funded by Cuba to serve as Salvadoran freedom fighters—the "Salvatruchas."[3] The union of the two gangs resulted in Mara Salvatrucha, a by-product of the assimilation process within the United States.

Mara Salvatrucha (MS) quickly gained recognition for its violence and business savvy. By maintaining ties to their homeland, the gang members became illegal arms dealers.[4] In El Salvador, according to Ortega, a hand grenade sells for $1; pen guns that fire single bullets and can be used to write ransom notes for $10; an M-16 rifle for $200; an AK-47 for $1,500. In the United States, these weapons can be sold from the trunks of cars for ten times the purchase price. It didn't require a Keynesian knowledge of advanced economics for the *jefes* to realize that a fortune could be made by purchasing such weapons and selling them to members of other street gangs throughout the United States.

Along with the weapons, MS became a major importer of illegal drugs. By muscle and machete, it managed to take over the established drug routes of other Latin gangs. By 1996, 70 percent of the cocaine and marijuana that flowed into the United States came from Mexico, and the lion's share of this trafficking remained under the control of MS.[5] Business became so brisk that the gang established warehouses for its illicit goods in Matamoros, Mexico, just south of Brownsville, Texas.[6]

Along with importing, MS soon engaged in exporting goods from the United States to Latin America—including massive amounts of baby formula, health and beauty supplies, cologne and perfume, diabetes test strips from Texas, electronics from Arizona and California, and thousands of automobiles from nearly every state in the Southwest.[7] By 1990, 80 percent of the cars on the streets of El Salvador, Guatemala, and Honduras had been stolen from the

United States by members of Mara Salvatrucha.[8] "We sell them cheap," Ortega boasts. "In Tegucigalpa, you can buy a new Cadillac for $15,000, sometimes less. A new Chevy sells for $3,000 to $4,000, cheap enough for an FBI agent to buy."

In addition to serving as a nefarious paradigm for NAFTA (North American Free Trade Agreement), the Salvadorans displayed the entrepreneurial spirit of their newly adopted country by taking over the methamphetamine ("crack") market from the Warlocks and other biker gangs.[9] "The Warlocks got high on their own supply and loafed around," Tim Carter, deputy sheriff of Shenandoah County told *New York Times* reporter Matthew Brzezinski. "What amazed me a lot about these guys [MS members] is that they had jobs. They'd put in 50 hours at the poultry plant and then drive down to the Carolinas to pick up loads of meth. They certainly weren't lazy."[10]

By 2005 Mara Salvatrucha expanded to Oregon, Texas, Nevada, Utah, Oklahoma, Illinois, Michigan, Maryland, Virginia, Georgia, Florida, New York, and even Alaska.[11]

Thousands of gang members came to settle in Washington, DC, and its surrounding environs in Maryland and Virginia. In 2003 more that seven hundred crimes related to gang activities were committed in the Shenandoah Valley of Virginia, a picturesque area once known for sun-bleached hayfields and wooden covered bridges.[12] In Fairfax County, where many families trace their ancestries to the antebellum era, a fourteen-year-old stabbed a local resident to death outside a feed store in order to impress his *homies*; a sixteen-year-old boy was attacked with machetes by MS thugs, who attempted to hack off his hands; and a seventeen-year-old girl was hacked to pieces by Denis Rabbit Raz, an MS *jefe* (boss), after she testified in court about the gang's activities.[13]

Other sections of the country, in the wake of MS's success and expansion, have witnessed similar horrors. In Klamath Falls, Oregon, for example, Maximiliano Silerio Esparza, a newly arrived gangbanger from El Salvador, raped two nuns who were praying on a walking path and then strangled one to death with her own rosary beads.[14] Thousands of MS members now occupy prison cells from Maine to California for such crimes as trafficking in illegal narcotics, home invasion, burglary, carjacking, extortion, witness intimidation, rape, and murders, including the assassinations of three federal agents.[15]

Originally, only Salvadorans could become members of Mara Salvatrucha. But the American experience soon taught the gang to set aside its parochialism and to band together with other groups from Central America. Cliques (as the various chapters of the gang are called) now boast members from Ecuador, Guatemala, Honduras, and Mexico. A few even include African Americans.[16]

Adherents to Mara Salvatrucha range in age from ten to forty. All sport tattoos, including the letters *MS* and the number *13*. The number refers to the initiation rite in which members beat gang "wannabes" with fists and clubs until they count to thirteen. Initiates must undergo the beating without a whimper or a word of protest, even though many are left with fractured jaws, broken arms, and cracked ribs.[17] Ortega lost his front teeth during his initiation into the gang in 1992.

Gang members, according to Ortega, share a common hand signal by forming the letter *M* with three fingers held within the palm and the hand pointed downward—the same gesture in the shape of a pitchfork that is employed by Italian peasants to invoke the evil eye.

Within the various cliques, members remain highly protective of their turf. They exact "taxes" on neighborhood shops and stores—even from the local prostitutes, who, in certain sections of Los Angeles, must cough up fifty per night for the right to stand on a street corner.[18]

In March 2004 Rockard Delgadillo, the Los Angeles city attorney, filed an injunction against Mara Salvatrucha, maintaining that the gang's criminal activities represent a "public nuisance" based on the number of its drug crimes, robberies, and killings.[19]

But Mara Salvatrucha represents far more than a growing public nuisance or a pressing sociological problem. It has come to pose a grave threat to national security. There is little doubt that another attack on America by al Qaeda looms in the immediate future—and this attack will have been made possible by this savage horde of barbarians.

Many migrants who want to cross the 1,820-mile border between Mexico and the United States now must come to terms with more than three thousand MS gang members who keep watch over this area like flocks of turkey vultures. The going rate for safe passage is five thousand dollars. Families without the required cost of admission for entrance into the magic kingdom of Madonna, Michael Jordan, and Mickey Mouse are required to relinquish all their possessions, including watches, family heirlooms, gold necklaces, silver bracelets, wedding rings, and even their shoes and sombreros. Some with neither money nor tangible goods must come up with other offerings, such as the sexual favors of their wives and children. Those who resist payment are often hacked to pieces or tossed from moving trains. "There are hundreds who are pushed off trains by the *maras* if they resist the robberies," says Asdrubal Aguilar Zepeda, the Salvadoran consul in Tapachula, Mexico, near the Guatemalan border. The bodies and body parts of dead migrants are often strewn along the railroad tracks from Tapachula to the northern regions of Mexico.[20]

Most illegal aliens continue to travel the well-established routes into the United States, including "cocaine alley," a passageway through the Sierra Madre that empties into Cochise County, Arizona. On any given day, one can see thousands making their way to Bisbee, Sierra Vista, Tombstone, and Nogales, where they wait in school yards, vacant buildings, and fast-food restaurants for rides to appear that will transport them via Route 10 to Tucson, Casa Grande, Phoenix, and southern California.[21] "People don't feel safe," says Jim Ellis, chief of police in Bisbee. "They [the illegals] come across in droves."[22]

More than 4,000 illegals enter the United States every day—enough to fill sixty commercial jumbo flights. This amounts to a total of 3 million a year.[23] Over 190,000 are from countries other than Mexico and have been dubbed "OTMs" ("other than Mexicans") by immigration officials.[24]

In 2003 Homeland Security estimated that between 8 and 10 million illegal aliens are living in the United States, and 400,000 have final orders of deportation but cannot be found. Eighty thousand of these "alien absconders" are convicted criminals who simply disappeared after receiving their prison sentences.[25] "Looking for these absconders in a nation of 293 million," Senator Jeff Sessions (R-AL) said, "is like looking for a needle in a haystack."[26]

By controlling leading passageways into the New World, Mara Salvatrucha has transmogrified into a multimillion-dollar enterprise, and, by so doing, it has become an international Mafia.

In the wake of 9/11, Mara Salvatrucha attracted the attention of top al Qaeda officials, who realized that the Salvadoran gang could be used to smuggle operatives and weapons (including small nukes and other weapons of mass destruction) into the United States[27] and also to provide safe transport over the border and shelter for sleeper agents in towns and cities from San Diego, California, to Bangor, Maine; from Miami, Florida, to Kodiak, Alaska—places where the gang already has established a presence.[28]

By 2002 the most notorious terrorist organization in the world came to a financial agreement with the most violent street gang on the American continent. The agreement was simple and straightforward. In exchange for safe passage across the border, shelter within the states, and a bogus *matricula consular*, al Qaeda—through its cells in South America—agreed to pay Mara Salvatrucha from $30,000 to $50,000 for each sleeper agent.[29]

Matricula consulars are official identification cards that are issued by the Mexican government through its consular offices. The cards verify that the bearers are Mexican citizens who are living outside of Mexico with the government's permission. They can be used by Mexican nationals to open bank accounts and obtain drivers licenses in the United States.[30] According to many US officials, these cards pose a serious threat to national security. Steve McCraw, assistant director of the FBI's Office of Intelligence, provided the following testimony to a House Judiciary Subcommittee on June 26, 2003: "The ability of foreign nationals to use the *matricula consular* provides an opportunity for terrorists to move freely within the United States without triggering name-based watch lists that are disseminated to local police officers. It also allows them to board planes without revealing their true identity. At least one individual of Middle Eastern descent has also been arrested in possession of the *matricula consular* card."[31]

Counterfeit *matricula consulars* are easy to come by. They sell for $90 on Alvarrado Boulevard and in MacArthur Park in Los Angeles.[32] In Times Square and outside Union Station, Ortega maintains, the cost for such bogus Mexican identification has now risen to $150.

News of the alliance between al Qaeda and Mara Salvatrucha prompted Honduran officials, including Security Minister Oscar Alvarez, to adopt a zero-tolerance law that makes membership in the street gang illegal and punishable by twelve years in prison. Members of Mara Salvatrucha responded to this legislation by beheading scores of victims and leaving notes on the bodies for the Honduran government. One note read: "Idiots, the end of the world is approaching."[33]

Between 2002 and 2004, thousands of so-called Special Interest Aliens (SIAs), with the help of Mara Salvatrucha "coyotes," made their way across the Mexican border and into the United States. They came from countries that pose national security concerns: Saudi Arabia, Syria, Iran, Pakistan, Afghanistan, Egypt, Somalia, Yemen, Jordan, Lebanon, and even Iraq. Overnight, cocaine alley became littered with discarded Muslim prayer blankets, pages from Islamic texts, and copies of Arabic newspapers. For this reason, law enforcement officials renamed the passageway as "terrorist alley" and a street leading north from the city of Douglas, Arizona, as "Arab Road."[34]

Along with the Special Interest Aliens from the Middle East, the Salvadoran gang also escorted hordes of Chinese nationals with an affiliation to al Qaeda across the border. In 2001 Secretary of Defense Donald Rumsfeld informed the press that a sizable number of Chinese nationals remained in alliance with bin Laden and joined the Chechens, Pakistanis, and Arabs in the defense of the

northern Afghan city of Kunduz.[35] The Chinese government added
that such nationals were Muslim Uighurs from the western Xinjiang
Province, many of whom had been trained at al Qaeda camps in
Pakistan and Afghanistan.[36] In 2005 several of these nationals were
arrested in Boston.[37]

In 2003 US border officials managed to apprehend 4,226 SIAs.
During the first nine months of 2004, that number rose to 6,022, an
increase of 42 percent.[38] Since space in detention cells remained at
a premium, the newly collared SIAs were simply released from custody as soon as they received their hearing dates from the immigration judges. Ninety-five percent failed to appear for their scheduled
hearings.[39]

When members of the border patrol and state and local law
enforcement officials complained of the wholesale release of SIAs—
some of whom may be plotting the next 9/11—into the general pop-
ulace, federal officials responded by insisting that such detainees
must be treated in the same way as detainees from countries that did
not harbor terrorists. Indeed, Asa Hutchinson, undersecretary of the
Office of Homeland Security, pointed out that the practice of
rounding up illegal aliens who appear to come from the Middle East
displays insensitivity and an attitude of racial profiling.[40]

And so the practice of arrest and release continued as code
orange gave way to code yellow. Alarms were raised from such
elected officials as Solomon P. Ortiz, the ranking Democrat on the
House Armed Services Subcommittee on Readiness, who confirmed
that Middle Easterners with possible ties to al Qaeda were being
detained only to be set free for lack of jail space. He said: "It's true.
It is very reliable information, from the horse's mouth, and it's happening all over the place. It's very, very scary, and members of Congress know about this. We have contacted several agencies, and I
have talked to some people, but I can't say who."[41]

Despite the voicing of such concerns, the Bush administration
instituted no measures to secure the Mexican border, and the situation was allowed to continue even after the arrest in April 2004 of
Mohammed Junair Babar, a leading al Qaeda operative, upon his
return to Queens, New York, after attending a terrorist summit in

Pakistan. Babar, who was born and raised in Queens, had been commissioned to carry out bombings in London on the model of the 1995 bombing of the Murrah Federal Building in Oklahoma City. Faced with seventy years in prison, Babar agreed to talk to US officials in exchange for entrance into the witness protection program. He said that a spectacular attack on American soil—the nuclear 9/11—was planned for the immediate future and that the key agents for this attack were being smuggled into the United States from Mexico.[42]

In June 2004 seventy-seven males "of Middle East descent" were arrested by patrol guards from Willicox, Arizona. The seventy-seven were trekking through the Chiricahua Mountains as part of a large caravan of migrant workers. "These guys didn't speak English," one field agent said, "and they were speaking to each other in Arabic. It's ridiculous we don't take this more seriously. We're told not to say a thing to the media." Another field agent added: "All the men had brand-new clothing and the exact same cut of mustache."[43]

Sheriff D'Wayne Jernigan of Del Rio, Texas, expressed his frustration at such developments by saying: "Are they criminals? Are they terrorists? We don't know who they are. The agency officials at this level here locally, I truly believe, are just as much against these releases as I am. They feel betrayed. They're thinking, 'We work hard to apprehend these people and then the next day someone at the Washington level orders their release.' Why are we apprehending them in the first place? They turn these people loose with a piece of paper that tells them to report to an immigration hearing at an unknown time and place and there's no way for an agency to get in touch with them again. Are they going to show up at those hearings? Will the agency ever be able to find them? Let's be realistic. It's ridiculous! A war on terrorism? Homeland security? Hah!"[44]

Asa Hutchinson eventually responded to such outcries by insisting that the American people lack the "will" to "tell our law enforcement people to go out there and uproot those 8 million here—some of whom might have been here 8 to 12 years, who got kids here that are American citizens—and to send them out of the country."[45]

In October 2004 US security officials received intelligence reports that a group of twenty-five Chechen terrorists with ties to al Qaeda may have entered the United States from Mexico. The report prompted the US Department of Education to issue a warning for school officials to examine the security measures since the Chechen Muslims were responsible for an attack on a school in Russia in which three hundred were killed and seven hundred wounded, most of whom were children.

The situation at the border now had veered from the absurd to the ridiculous. To illustrate this point, members of the America Border Patrol crossed into Mexico, using a well-traveled route, and brought back a simulated weapon of mass destruction in a back-pack. They delivered the simulated weapon to the steps of the Tucson Federal Building. Mike King, a member of the group, said that he hoped the experiment would convince the Bush administration that the country's southern border is a national security risk. King's hope for attention was in vain. Throughout the summer of 2004, the border remained as porous as before with illegals—including many suspicious figures from the Middle East—crossing into the United States as though they were participating in a gala parade. This prompted Mike King's group to perform the experiment several times, again with similar success. Never did any law enforcement officer stop to question members of the group about the "weapon" as they crossed the border

On December 4, 2004, Frankie Sanchez-Solorzano, an MS-13 member, was arrested while guiding a group of illegals across the Rio Grande. One member of the group was Fahrhue Islam, a Bangladeshi with ties to Sunni terrorist groups. This incident, which attracted scant national press coverage, nevertheless, raised new concerns among public officials. Although Bangladesh was not listed as a nation that sponsors terror, the predominantly Muslim south-central Asian nation remained a "country of concern" for US intelligence. "It is a very poor country where many people earn as little as sixty-five cents a day," Congressman Solomon Ortiz said. "The question I'm asking is, 'How was he able to finance his way here?'"[46] No answer was forthcoming. The mysterious man from Bangladesh was released from custody upon being granted a date for a hearing before an immigration judge and promptly disappeared within the proverbial melting pot.[47] The story was not without a denouement. Sanchez-Solorzano, the MS guide, turned out to be wanted for the murders of two police officers in Honduras.[48]

Any further doubts about Mara Salvatrucha and the Mexican border were put to rest when Elmer "Tiger" Tejada was spotted in East Boston several days before Christmas in 2004. The previous January, Tejada had been deported by Boston Homeland Security Immigration and Customs Enforcement officials after being convicted of a host of crimes, including attempted murder for hurling a machete at Chelsea cops.[49]

Most alarming was the report that Adnan el-Shukrijumah, the appointed Mohammad Atta for the next 9/11, had secured the services of Mara Salvatrucha to reenter the United States after a series of all-out alarms had been issued for his arrest.

Adnan el-Shukrijumah was born in Guyana on August 4, 1975—the first born of Gulshair el-Shukrijumah, a forty-four-year-old radical Muslim cleric, and his sixteen year old wife. In 1985 Gulshair migrated to the United States, where he assumed duties as the *imam* of the Farouq mosque—a mosque that raised millions for the *jihad* and that served as a recruiting station for al Qaeda.[50] In 1995 the Shukrijumah family moved to Miramar, Florida, where Gulshair became the spiritual leader of the radical Masjid al-Hijah and where Adnan became friends with José Padilla, who planned to detonate a radiological bomb in midtown Manhattan; Mandhai Jokhan, who was convicted of attempting to blow up nuclear power plants in southern Florida; and a group of other home-grown terrorists.[51]

Adnan attended flight schools in Florida and Norman, Oklahoma, along with Mohammad Atta and the other 9/11 operatives, and he became an accomplished commercial jet pilot, although he never applied for a license with the Federal Aviation Commission.[52]

In April 2001 Adnan spent ten days in Panama, where he reportedly met with al Qaeda officials to assist in the planning of 9/11. He also traveled to Trinidad and Guyana. The following month, he obtained an associate's degree in computer engineering from Broward Community College.[53]

During this time, he managed to amass passports from Guyana, Trinidad, Saudi Arabia, the United States, and Canada. He also began to adopt a number of aliases, including Adnan G. el-Shukri-jumah, Abu Arifi, Jafar al-Tayyar, Jaafar At Yayyar, Ja'far al-Tayar, and Mohammed Sher Mohammed Khan (the name that appeared on his official FBI file).[54] He traveled to Saudi Arabia and Pakistan, where he met with Ramzi Binalshibh, Khalid Shaikh Mohammed, and other members of the al Qaeda high command.[55] He also spent considerable time within al Qaeda camps in Afghanistan, where he was trained in explosives.[56]

El-Shukrijumah proved to be unique among the terrorists. Not only could he speak English without the slightest trace of an accent, but he also possessed the ability to blend into a crowd, to alter his looks, and to assume a multitude of identities. He represented the proverbial Mr. Cellophane.

Following the success of 9/11, el-Shukrijumah became singled out by bin Laden and al-Zawahiri to serve as the field commander for the next great attack on US soil—the so-called American Hiroshima—that would leave millions dead and the richest and most powerful nation on earth in ashes. In preparation for this mission, Adnan, along with fellow al Qaeda sleeper agents Anas al-Liby, Jaber A. Elbaneh, and Amer el-Maati, was sent to McMaster University in Hamilton, Ontario, a facility that boasted a five-megawatt nuclear research reactor, the largest reactor of any educational facility in Canada.[57]

At McMaster, where they may have enrolled under aliases, el-Shukrijumah and his associates wasted little time in gaining access to the nuclear reactor and stealing more than 180 pounds of nuclear waste for the creation of radiological bombs.[58]

In the wake of Operation Enduring Freedom, US military officials discovered the reoccurrence of el-Shukrijumah's name in "pocket litter"—documents and scraps taken from prisoners and dead al Qaeda soldiers.[59] The name did not set off immediate

alarms since Khalid Shaikh Mohammed, al Qaeda's military operations chief, was captured in Karachi, Pakistan. After days of interrogations, coupled with severe sleep deprivation, Khalid Mohammed,
as noted earlier, told US officials that bin Laden was planning to
create a "nuclear hell storm" in America.[60] Unlike other attacks, the
terrorist chief said, the chain of command for the nuclear attack
answered directly to bin Laden, al-Zawahiri, and a mysterious scientist called "Dr. X."[61] He further confessed that the field commander
was a naturalized American citizen whom he called "Jafer al-Tayyar"
("Jafer the Pilot").[62]

US officials soon learned that Jafer the Pilot was an alias for
Adnan el-Shukrijumah and that he and other al Qaeda agents had
been sent to McMaster. By the time Canadian officials became
alerted to the situation, it was too late—the elusive Adnan had
already crossed the border into Buffalo, where he remained sheltered by members of a local mosque.[63]

On March 21, 2004, Attorney General John Ashcroft and FBI
director Robert Mueller issued a BOLO ("be-on-the-lookout") for el-
Shukrijumah and Amer el-Maati, along with Aafia Siddiqui, a Pakistani woman who received a biology degree from MIT and penned
a doctoral thesis on neurological science at Brandeis; Ahmed
Khalfan Ghailani (aka "Foopie"), who took part in the 1998
embassy bombings in Kenya and Tanzania; and Adam Yahiye
Gadahn (aka Adam Pearlman), a convert to Islam who grew up on
a goat ranch in Riverside County, California. Also mentioned was
Abderraouf Jdey, the leader of the al Qaeda cell in Toronto.[64]

Several days after the BOLO was released, Adnan and Jdey were
spotted at a Denny's Restaurant in Avon, Colorado, where one
ordered a chicken sandwich and a salad. The restaurant manager
described them as "demanding, rude, and obnoxious."[65]

Following this sighting, the diminutive Adnan, who stands at
five feet four and weighs 132 pounds, resurfaced at a terrorist
summit in the lawless Waziristan Province of Pakistan in April 2004.
The summit has been described by the FBI as a "pivotal planning
session" in much the same manner a 2000 meeting in Kuala
Lumpur was for the 9/11 attacks. Attending the meeting were

Mohammed Babar, who (as stated above) has been charged with buying materials to build bombs for use in attacks in Great Britain, and al-Hindi, a Pakistani technician whose computer contained plans for staging attacks at financial institutions in New York, New Jersey, and Washington, DC.[66]

On May 27, 2004, el-Shukrijumah was spotted at an Internet café in Tegucigalpa, the hilly capital of Honduras, where he made calls to France, Canada, and the United States.[67] He was described as badly dressed and bearded. At his table were Mara Salvatrucha leaders (*jefes*) from Panama, El Salvador, Mexico, and Honduras.[68]

From Tegucigalpa, he made his way north to Belize in British Honduras and, from Belize, to Mexico's Quintana Roo State, south of Cancun.[69]

El-Shukrijumah remained in Mexico for much of the summer. In late August, he was spotted in the northern Mexican province of Sonora near "terrorist alley."[70] Alberto Chapetti of the US Consulate in Nogales issued warnings about the presence of the terrorist in the area along with the posting of a $5 million reward for tips leading to his arrest.[71]

Concern about el-Shukrijumah's sojourn in Mexico was heightened in November 2004 with the arrest in Pakistan of Sharif al-Masri, a key al Qaeda operative. Al-Masri, an Egyptian national with close ties to Ayman al-Zawahiri, informed interrogators that al Qaeda has made arrangements to smuggle nuclear supplies and small tactical nuclear weapons into Mexico. From Mexico, the weapons were to be transported across the border with the help of a Latino street gang.[72]

In response to this information, US officials began monitoring all heavy trucks crossing the border, while Mexican officials pledged to keep close watch over flight schools and aviation facilities. Such precautions may have been adopted too late. A Piper PA Pawnee crop duster had been stolen from Ejido Queretaro near Mexicali on November 1, 2004. The plane's tail number was XBCYP. The thieves, Mexican officials surmised, were either drug dealers or al Qaeda operatives, and clearly one was a highly trained pilot.[73]

Ortega believes that el-Shukrijumah has already crossed the border, that plans for the next 9/11 are well under way, and that it remains too late for US officials to take appropriate action to prevent the coming nuclear nightmare. He claims that he has transported scores of illegals from the Middle East who could not speak a word of Spanish across the border. "These people are not coming here to pick vegetables, mow lawns or pluck chickens," he says. "They have something else in mind."

Asked why members of Mara Salvatrucha would agree to aid al Qaeda in the wake of 9/11, Oretga says: "Money. That is the reason for everything."

Ortega butts his cigarette, rises from the table, and walks to the men's room. Customers stare at the muscular figure as he walks past them. To some, he must appear as the crystallization of a fabulous character from literature—Melville's Queequeg or Bradbury's Illustrated Man. To others, he must seem to be a sideshow figure—the tattooed man on a leave from some traveling circus.

No one in the crowded restaurant can see the initials that have been engraved over the broad expanse of Ortega's back. They remain masked by his white, sleeveless muscle shirt. They are sported on the bodies of almost all members of Mara Salvatrucha since they represent the guiding principle of the Salvadoran street gang.

The initials F. T. W.

F. T. W. stands for "fuck the world."

CHAPTER ELEVEN
The Sleeper Cells

*We have to be terrorists. . . . The great Allah said, "Against them make
ready your strength to the utmost of your power including steeds of war, to
strike terror [into the hearts of] the enemies of Allah and your enemies."*
—Sheikh Omar Abel Rahman, speaking at a
mosque in Los Angeles, December 1992

*I have a vision in America, Muslims owning property all over, Muslim
businesses, factories, hatal meat, supermarkets, all these buildings owned
by Muslims. Can you see the vision? Can you see the Newark International Airport and the John Kennedy Airport and the LaGuardia having
Muslim fleets of planes and Muslim pilots? Can you see our trucks rolling
down the highways [with] Muslim names? Can you imagine walking
down the streets of Teaneck [New Jersey] and seeing three Muslim high
schools, five Muslim junior high schools, and fifteen [Muslim elemen-
tary] schools? Can you see the vision? Can you see young women
walking down the streets of Newark, New Jersey, with long flowing hijab
and long dresses? Can you see the vision of an area of no crime, controlled by the Muslims?*
—Siraj Wahhaj, the first Muslim to deliver the daily prayer
invocation for the US House of Representatives, 1991

Knowing what I know, I can confidently say that until the investigative responsibilities for terrorism are transferred from the FBI, I will not feel safe. . . . The FBI has proven for the past decade it cannot identify and prevent acts of terrorism against the United States and its citizens at home and abroad. Even worse, there is virtually no effort on the part of the FBI's International Terrorism Unit to neutralize known and suspected international terrorists living in the United States.
—FBI agent Robert Wright, May 30, 2002

When you arrive at the 500 block of Atlantic Avenue in Brooklyn, New York, you might believe that you have passed through a space warp and landed in Cairo. The avenue is lined with dozens of exotic stores and crowded shops featuring such items as prayer mats, *misbahahs* (Muslim "rosaries"), Islamic books, Egyptian fragrances, Wahhab clothing (including the latest in burqas), *Nikah* and *Aqiqah* greeting cards, Middle Eastern groceries, traditional Muslim headdress (*hijabs* and turbans). The restaurants offer such delicacies as roasted goat with Sudanese sauce and *tikkas* (spiced grilled mutton) with *naap* (flat bread) and imported pomegranates. The shop owners and the local customers are strangely subdued, as if obeying a dictate of decorum that remains unknown to intruders from Manhattan and other sections of Brooklyn.

The Farouq Mosque at 554 Atlantic Avenue, one of the 130 mosques in New York City, does not look like a mosque. It is a bleak storefront that resembles the poorly lit, private clubs in Little Italy, where wiseguys gather on Saturday afternoons to shoot pool or play cards. The only indication that this property is a place of worship comes from the Arabic words above the entranceway: "*Ashhadu an la ilaha ilaha illa Llah, wa ashhadu anna Mohammad rasulu Llah*" ("There is no God but Allah, and Muhammad is his prophet").

If you enter the building, you are likely to come face-to-face with several large, bearded, serious African Americans, Arabs, or Yemenis in robes and turbans. They serve to protect the *musallah* (main prayer room) on the third floor and other sections of the mosque (the meeting room and the *madrassah* on the second floor) from curious intruders, government officials, and journalists. If you attempt to make your way to the stairwell or the elevator, you will be halted. "You do not belong here," a guard might say, while jabbing his index finger at your solar plexus for emphasis. The fact that you don't belong here is clearly discernible by your failure to recite the proscribed greeting (*Allahu Akbar*), your lack of the proper attire including a headdress, and your neglect to remove your shoes.

If the sun is beginning to set, flag down a cab and be happy to get out of this section of Brooklyn and get home safely.

Established in 1976, the Farouq Mosque may be one of the most dangerous places in the United States. Its principal function, over the years, has not been to bring one to grips with a transcendent being named Allah but rather to raise money for the *jihad*. Mohammed Ali Hasan al-Moayad, a fund-raiser from the mosque, personally delivered $20 million to bin Laden and another gift of $3.5 million to the Al-Aqsa Society, the financial arm of HAMAS.[1] This represented the amount that Moayad and his fellow fundraisers had amassed in December 1999.[2] How much more money has been raised for radical Islam from this inconspicuous storefront in Brooklyn remains anyone's guess.

The mosque has been responsible for sending recruits to al Qaeda training cells throughout the Middle East.[3] One such recruit was Jamal Ahmed al-Fadl, who worked to secure nuclear weapons and supplies for al-Qaeda, during bin Laden's sojourn in Sudan.

The Farouq Mosque first came to prominence on November 5, 1990, when El-Sayyid Nosair, a leading member, murdered Rabbi Meir Kahane, the founder of the Jewish Defense League, in the ballroom of the Marriott East Side Hotel, where Kahane was speaking. Investigators discovered that Nosair, along with Nidal Ayyad, Clement Hampton-El, Mohammed Salameh, and other members of the mosque were being trained for the *jihad* with rifles, shotguns, 9-mm and .357 caliber handguns, and AK-47 assault weapons at the Calverton Shooting Range on eastern Long Island by Ali Mohamed, a sergeant at Fort Bragg and an al Qaeda sleeper agent.[4]

The killing of Kahane represented the first violent outburst of radical Islam on American soil. In his notebook, Nosair expressed his vision of the holy war within the land of the Great Satan (the United States) by calling for "the breaking and destruction of the enemies of Allah. And this is by means of destroying, exploding, the structure of their civilized pillars such as the tourist infrastructure which they and their high world buildings which they are proud of and their statues which they endear and the buildings which gather their head[s], their leaders, and without any announcement for our responsibility of Muslims for what had been done."[5] In 1990 Nosair, employed as an air-conditioning repairman for the city of New York, had penned the idea for a defining moment of radical Islam: the terrorist attacks that were to occur on September 11, 2001. Few within the FBI and the CIA read the notebook. It had been written in Arabic and was not translated by US officials until years later.[6] This lapse in security may serve to explain why intelligence officials remained oblivious to the situation that was brewing within the Brooklyn mosque.

The mosque, at this time, served as the headquarters of the Al-Kifah Refugee Center, a purported front for al Qaeda, with branches in Atlanta, Boston, and Tucson, and recruiting stations in a total of twenty-six states.[7] Founded by bin Laden's mentor Abdullah Azzam, the Al-Kifah Refugee Center became identified by an aide to Egyptian president Hosni Mubarak as a CIA front for transferring funds, weapons, and recruits to the anti-Soviet *mujahadeen* in Afghanistan. It became part of the Brooklyn mosque in 1986 and began receiving over $2 million a year from the Reagan administration.[8] Azzam became a frequent visitor to Brooklyn. He can be seen and heard in a 1988 videotape, telling a large crowd at the mosque that "blood and martyrdom are the only way to create a Muslim society."[9]

By 1992 the Al-Kifah Refugee Center became a haven for veterans from the *jihad* in Afghanistan, who had entered the country with passports arranged by the CIA.[10] A feud between the newcomers and the older members of the Farouq Mosque erupted, which may have resulted in the murder of Mustafa Shalabi, the fiery *imam*, on March 1, 1991.[11] The murder remains unsolved.

Shalabi's replacement was Sheikh Omar Abdel Rahman, the

revered blind cleric who had provided the religious authorization for the assassination of Egyptian president Anwar Sadat.[12] Sheik Omar had migrated to Brooklyn in July 1990, after the consular section of the US embassy mistakenly granted him a visa, even though his name appeared on a high-alert watch list. The new leader of the Al-Kifah Refugee Center became Wadith el-Hage, who arrived in Brooklyn from Tucson the day after Shalabi's murder. El-Hage, a naturalized American citizen, later became the personal secretary of Osama bin Laden.[13] Pieces were now in place for terrorist attacks within the belly of the "Great Satan" (the blind sheikh's favored name for the United States of America).

The first great attack on American soil came with the bombing of the World Trade Center on February 26, 1993, which killed six people and injured 1,042, producing more hospital casualties than any other event in domestic American history apart from the Civil War.[14] Attention was again attracted to the Farouq Mosque, where three of the terrorists (Mohammed Salameh, Mahmud Abouhalima, and Nidal Ayyad) involved in the bombing had been prominent members and where the plans for the attack had been hatched by Ramzi Yousef, a Kuwaiti radical who had been educated in Great Britain. Yousef had arrived in New York in August 1992 with an Iraqi passport and a desire to kill 250,000 American citizens. The number, Yousef would later explain, was not random. It represented the number that had been killed by the atomic bombs that had been dropped on Hiroshima and Nagaski at the close of World War II.[15]

Knowing that the Brooklyn storefront had become a hive for terrorist activities, Neil Herman, a senior FBI agent, asked the Justice Department for permission to wiretap the mosque and the *imam*'s office. The request was denied because the address represented a "house of worship."[16]

Within the house of worship, the blind sheikh was preaching *jihad* and calling upon his followers to fight the Great Satan with bullets and bombs as in this sermon from 1993:

> There is no solution for our problems except *jihad* for the sake of
> God. . . . There's no solution, there's no treatment, there's no med-
> icine, there's no cure except with what was brought by the Islamic

method which is *jihad* for the sake of God. . . . No, if those who have the right to have something are terrorists then we are terrorists. And we welcome being terrorists. And we do not deny this charge to ourselves. And the Koran makes it, terrorism, among the means to perform *jihad* in the sake of Allah, which is to terrorize the enemies of God and who are our enemies, too. . . . They say that he who has done his job during the day in order to go to the mosque has performed *jihad*. And he who listens to a religious lecture has performed *jihad*. What is this? This is distortion to the subject of *jihad*. Praying, listening, *jihad*? Coming to the mosques is a good work. And group praying is just praying. *Jihad* is fighting the enemies. Fighting the enemies for God's sake in order to raise them high in his word. . . . We don't fight the enemies unless we have guns, tanks and airplanes equal to those of the Soviet Union.[17]

Following the World Trade Center bombing, radical members of the mosque sought new ways to mount another attack. They allegedly came up with a plan to bomb the Lincoln and Holland tunnels, the United Nations, the FBI headquarters, and various federal office buildings in Manhattan. To create the bombs, they began to mix Scott's Super Turf Builder with 255 gallons of diesel fuel within a safe house in Queens.[18] Neighbors noted the strange comings and goings of men in robes and turbans. A tip was placed to FBI headquarters, resulting in a raid on June 23, 1993. Eleven men, including blind Sheikh Omar, were collared and charged as coconspirators in a terrorist plot.

Gulshair el-Shukrijumah, another imam at the Farouq Mosque, served as Sheikh Omar's interpreter at the trial. He also acted as a character witness for Clement Rodney Hampton-El (aka, Dr. Rashid), a member of the congregation.[19] After the trial, Gulshair and his family migrated to Miramar, Florida. His son Adnan el-Shukrijumah has been identified by the FBI as the next Mohammad Atta and remains at large as the most wanted terrorist in North America.[20]

In 2003 Amin Awad, the new *imam*, was prevented from counseling prisoners in New York's Rikers Island prison because of his mosque's affiliation with al Qaeda and international terrorism.[21]

Less than twenty minutes from the Farouq Mosque is the Al-Salaam Mosque in Jersey City. It is a squalid storefront in the midst of a ghetto area. The blind sheikh also served as the imam at this place. Several members of the congregation were involved in the 1993 World Trade Center bombing as well as the sheikh's grandiose plans to create destruction throughout Manhattan.[22]

There is a warning above the doorway to this mosque that reads: "Those who do not belong to this Muslim community will be persecuted for trespassing."[23] Make no mistake about it. You are not welcome here, either, as the guards that stand at the doorway, no doubt, will tell you in no uncertain terms. Before you are expelled, you might have time to read the notice on the bulletin board. It informs Muslims who worship here that they are not required to reveal their immigration status to the police or speak to inquiring officials from the US Immigration and Naturalization Service or to respond to questions from law enforcement officials. If the FBI threatens to get a grand jury subpoena, do not submit, the sign says.[24]

The Brooklyn and Jersey City storefront mosques are not the only cases within the United States. On January 7, 1999, Shaykh Hisham Kabbani, chairman of the Islamic Supreme Council of America, as noted earlier, expressed his belief before a committee of the US Department of State that 80 percent of the mosques in America have been taken over by Muslim extremists who are committed to the destruction of the United States.[25] Kabbani went on to say that Muslim schools, youth groups, community centers, political organizations, professional associations, and commercial enterprises within the United States share the views of Islamic militant groups, such as al Qaeda, and remain extremely hostile to American culture, wanting to replace it with an Islamic order.[26] According to the *Christian Science Monitor*, Rep. Paul King of New York, a member of the House Select Committee on Homeland Security, maintained that Shaykh Kabbani's 1998 assessment of militant Islam within the United States was too conservative, arguing that 85 percent of the mosques have radical leadership and that members of these mosques are not cooperative with law enforcement in the war on terror.[27]

Of the twelve hundred mosques in America, 80 percent have been built with money from the Sunnis in Saudi Arabia; the remaining 20 percent from the Shiites in Iran. The Saudis have spent more than $87 billion since 1973 to spread the Islamic faith in the Western Hemisphere. Of that amount $70 billion has gone for mosque construction in America, including the Islamic Cultural Center on the Upper East Side in New York and centers of Islamic worship in Los Angeles, Washington, Chicago, Maryland, Ohio, and Virginia.[28] "It is very difficult for American Muslims to collect enough money to build their own mosques and so they rely on these [Saudi Arabian] institutions," Ahmed al-Rahim, a professor of Arabic Language and Literature at Harvard University, told an Ethnic and Public Center panel in Washington, DC, in 2003. "That reliance brings political baggage with it, as mosques become subservient to an entire political program, a very radical political program."[29] Since the Sunnis of Saudi Arabia are predominantly Wahhabists, a large number of US mosques reflect Wahhabist beliefs and practices.

Since a number of mosques throughout the United States remain hotbeds for radical Islam, it is easy for al Qaeda sleeper agents, who have been shepherded over the border by Mara Salvatrucha, to find refuge in any major metropolitan area. They simply have to appear at a receptive Muslim community center or a receptive Islamic place of worship. It is small wonder, therefore, that America remains faced with nests of al Qaeda agents from Maine to California. In his 2002 State of the Union address, President George W. Bush warned of the proliferation of al Qaeda sleeper cells and agents by saying: "Thousands of dangerous killers, schooled in the methods of murder, often supported by outlaw regimes, are now spread throughout the world, set to go off without warning."[30]

Sleeper agents represent an integral part of the al Qaeda network, a network that "has always been divided into two halves."[31] One half is now based in the lawless frontier provinces of Pakistan, where Osama bin Laden and Ayman al-Zawahiri remain in hiding, and trouble spots throughout the Middle East, including Iran, Iraq, Syria, Lebanon, Saudi Arabia, and Sudan. This half, according to US military officials, will be eliminated during the course of the war on terror. The other half is infinitely more troublesome. Tens of thousands of al Qaeda agents, trained in Afghanistan, have settled into more than fifty countries throughout the world, where they are preparing further acts of terror to perpetuate the *jihad* long after the demise of al Qaeda's current leadership.[32] Documents discovered in Afghanistan point to much larger numbers of sleepers than previously imagined by intelligence sources. Some say that there are more than five thousand in North America, preparing for the "American Hiroshima." This figure was upheld by Shaykh Kabbani in his testimony before the US Department of State.[33]

Islamic terrorist cells can be found in every major American city. In New York, there are cells of al Qaeda, al Muhajiroun, al-Gama'a al-Islamiyya, Hezbollah, HAMAS, Fatah, and the PLO; in Philadelphia, al Qaeda, al Muhajiroun, al-Gama'a al-Islamiyya, Hezbollah, HAMAS, and the PLO; in Boston, al Qaeda and HAMAS; in Washington, DC, Hezbollah, HAMAS, al Qaeda, and the Palestinian Islamic Jihad;[34] in Raleigh, North Carolina, HAMAS and the Palestinian Islamic Jihad; in Miami, al Qaeda, HAMAS, Fatah, Palestinian Islamic Jihad, and the PLO; in Detroit, al-Gama'a al-Islamiyya, Hezbollah, HAMAS, al Qaeda, Fatah, and the PLO; in Houston, al Qaeda, HAMAS, Muslim Brotherhood, and the Palestinian Islamic Jihad; in San Francisco, al Qaeda, HAMAS, and the Abu Sayyaf Group; and in Los Angeles, HAMAS, al-Gama'a al-Islamiyya, and al Qaeda.[35] There are terrorist cells within Newark, Trenton, Atlanta, Seattle, Portland, Cincinnati, Cleveland, St. Louis, Indianapolis, Las Vegas, San Diego, Houston, Dallas, Denver, New Orleans, and even Norman, Oklahoma.[36] This is only a partial list. As of this writing, there remains not a single section or region of the country that remains free from terrorist infiltration.[37]

Most of the al Qaeda, Abu Sayyaf, al-Gama'a al-Islamiyya, al Muhajiroun, Muslim Brotherhood, and Fatah sleeper agents who are now within the United States, have been trained at al Qaeda camps in Afghanistan before 9/11. Many traveled to these camps from *madrassahs* throughout the Middle East (including the six thousand in Pakistan). Others were recruited from mosques and Muslim schools in Indonesia, Europe, and the United States. Most candidates were young men between the ages of eighteen and twenty-four who had been injured by ethnic conflict or who came from the lower levels of society.

At the camps, the recruits learned how to handle semiautomatic rifles (especially the Russian-made Kalashnikovs, the favorite weapon of bin Laden and his supporters of Afghani-Arabs), to pick planes out of the sky with surface-to-air missiles (including the Stingers that had been provided to the *mujahadeen* in its struggle against the Soviets), and to pack plastic explosives within suitcases (a technique that had been developed by the CIA). They also became subjected to intense religious training, during which they were obliged to commit long passages of the Koran to memory. At the conclusion of the six to eight weeks of basic *jihad* training, the recruits were compelled to swear a blood oath of allegiance, known as the *bayat*, to Osama bin Laden.

Zacarias Moussaoui, the first person indicted for the 9/11 attacks, offered the following testimony to the importance of the Afghan camps in preparing terrorists for the *jihad*:

> Once in the camp, it is easy, as in any sect, to make him [the recruit] lose his bearings. First of all, he is put through athletic training and then training in weapons handling. These are intensive exercises. He is always being set challenges that are increasingly difficult to meet. The young recruit is not well fed. He gradually becomes exhausted. He never manages to completely come up with what is being asked of him. After several weeks, he gets the feeling that he's not capable of doing what is expected of him. He experiences a feeling of embarrassment and malaise. In his own eyes, he is completely belittled: he feels guilty because he is incompetent. And yet he is told over and over again that others before him have succeeded and gone on to great things. . . . And if he carries on, it is to the bitter end. Because the only thing he can do to help the cause is to give his life to it. And this will also prove to

others that, at the end, he met their expectations. He is now ripe for suicide.[38]

After the *bayat*, select recruits (usually those well educated and from middle-class families) become chosen for "special operations training" at one of al Qaeda's top-level *jihad* camps, such as the Zahwar Kal al-Bar camp in the Kush Mountains. Here they receive training in foreign languages, intelligence management, kidnapping, hijacking, explosives, and methods of mass murder. The *Al Qaeda Training Manual*, a copy of which was found in Manchester, England, on May 10, 2000, outlines the following guidelines of deportment for sleeper agents in a country of *kafirs*, such as Great Britain and the United States:

1. Keep the passport in a safe.
2. All documents of an undercover brother, such as identity cards and passports, should be falsified.
3. When the undercover brother is traveling with a certain identity card or passport, he should know all pertinent information, such as his name, profession, and place of residence.
4. The brother who has special work status, such as a commander or communications link, should have more than one identity card or passport. He should learn the contents of each, the nature of the indicated profession, and the dialect of the resident area listed in the document.
5. The photograph of the brother in these documents should be without a beard. It is preferable that the brother's public photograph on these documents also be without a beard.
6. When using identity documents in different names, no more than one such document should be carried at one time.[39]

The manual goes on to state the series of precautions that must be observed by all agents when eating, drinking, applying for employment, and seeking living quarters within the country of nonbelievers. It offers the following guidelines for using public transportation:

1. One should seek public transportation that is not subject to frequent checking along the way, such as crowded trains or public buses.
2. Boarding should be done at secondary stations, since main stations undergo more careful surveillance.
3. The cover should match the general appearance—tourist class, first class, second class, and so on.
4. The document used should support the cover.
5. Important luggage should be placed among the passengers' luggage without identification tags. If it is discovered, its owner will not be arrested.
6. The brother traveling on a "special mission" should not arrive in the destination country at night because then travelers are few and there are search parties and checkpoints along the way.
7. When cabs are used, conversation of any kind should not be started because many cab drivers work for security companies.
8. The brother should exercise extreme caution and apply all security measures to the other members.[40]

In addition to the above, the manual specifies tried-and-true methods of appearing like the "polytheists" in Europe and the United States in order to escape suspicion. These methods range from carrying rosary beads to taking an unbeliever as a wife. Other helpful hints are as follows:

1. Never reveal your true name to cell members or associates at your place of employment.
2. Have a general appearance that does not indicate Islamic orientation, such as a beard, a long shirt, or even a toothpick.
3. Maintain friendly relationships with family members, if married to a *kafir*, and neighbors.
4. Never speak loudly.
5. Never get involved in advocating good or denouncing evil.
6. Never park in a no-parking zone or take photographs where it is forbidden.

7. Never undergo a change in your daily routine.
8. Never talk to your wife about *jihad* work.[41]

The closing sections of the terror manual contain the following instructions regarding the selection of a proper place to store weapons of mass destruction, such as nuclear suitcase bombs:

1. The arsenal should not be in a well-protected area or close to parks or public places.
2. The arsenal should not be in "no-man's-land."
3. The arsenal should not be in an apartment previously used for suspicious activities.
4. The place selected should be purchased by the agent or rented on a long-term basis.
5. The brother responsible for storage should not visit the arsenal frequently nor toy with the weapons.
6. Only the weapons keeper and the commander should know the location of the arsenal.[42]

Al Qaeda sleeper agents may now be found not only within some American mosques, colleges, banks, businesses, governmental offices, social agencies, but even within the inner ranks of the armed forces, the national media, and the intelligence community. For starters, consider the case of Ali Abdelsoud Mohamed. Mohamed enlisted in the US Army for a three-year stint in 1986. His first assignment was to a supply company at Fort Bragg, North Carolina—the headquarters of the Army's Special Forces: the Green Berets. Ali gained entry to the Special Forces and, within a year, rose to the rank of supply sergeant. During the Persian Gulf War of 1991, the Green Berets went deep into Iraq, where they guided US bombers to their targets. In 1993 they were deployed to Afghanistan to hunt down bin Laden and to locate the cells of al Qaeda.[43] For this reason, Fort Bragg represents the last place where one would expect to find a sleeper agent, and Mohamed's ready acceptance into the army's most elite and secretive corps provides vivid proof of al Qaeda's amazing ability to stretch its tentacles into the very heart of military intelligence.[44]

Mohamed was born in Egypt in 1952. He attended a military academy in Cairo and, upon graduation, entered the Egyptian army, where he spent thirteen years, rising to the rank of major. In 1984 he was dismissed from duty because of his affiliation with Egypt's Islamic *Jihad*, the group responsible for the assassination of President Anwar Sadat.[45]

In 1985 Mohamed moved to Sacramento, California, and married Linda Sanchez, an American medical technician. The following year, he enlisted in the US Army and gained full American citizenship, a valued prize for an al Qaeda agent.

In 1989 Mohamed received an honorable discharge and began traveling back and forth to Afghanistan, where he provided training in special operations (the same training he had received as a Green Beret) to recruits at al Qaeda training camps.

In 1991, he was recruited by the Al-Kifah Refugee Center at the Farouq Mosque in Brooklyn to provide training in military tactics for those members who would become involved in the 1993 bombing of the World Trade Center. That same year found Mohamed in Sudan, where he assisted bin Laden in establishing a new base of operations and where he trained al Qaeda operatives in intelligence techniques and the proper means of detonating explosive devices (the same techniques he had learned at Fort Bragg).[46]

When he returned to the United States, Mohamed brought Ayman al-Zawahiri with him to launch a massive fund-raising campaign for the *mujahadeen*.[47]

In the midst of all this activity, Mohamed was questioned by the FBI after his name and address were found on an al Qaeda agent who was attempting to enter the United States from Vancouver. During his 1993 questioning by FBI officials, he openly admitted that he was a member of bin Laden's terrorist group and that he had trained hundreds of al Qaeda operatives at training camps in Afghanistan and Sudan. Despite this startling and open admission of sedition, treason, and terrorism, the FBI released

Mohamed from custody because he had passed a carelessly administered lie-detector test.[48]

During the next four years, Mohamed made a series of trips to Afghanistan to assist in the establishment of training camps and to Kenya and Tanzania to make preparations for the bombing of the US embassies—an event that took place on August 7, 1998, killing 234 people, including twelve Americans, and wounding over five thousand others.[49]

The day after the bombings, FBI agents arrested Mohamed at his Sacramento home with a warrant from a special judge under the Federal Intelligence Surveillance Act. Within Mohamed's place of residence, the agents found charts for blowing up buildings and bridges throughout the United States and encoded messages regarding al Qaeda's plans for the future.[50]

After pleading guilty (yet again) to his crimes, Mohamed informed federal officials that their current profiles of sleeper agents were out of date.[51] Even the great Paddy Chayevsky (author of such film satires as *The Hospital* and *Network*) would be hard pressed to conceive of a more egregious, fictional example of the failure of the US military and intelligence superstructure to spot a sleeper agent than the incredible but true story of Ali Abdelsoud Mohamed. As Nabil Sharef, a former Egyptian intelligence officer, observed: "For [more than] five years, he [Mohamed] was moving back and forth between the United States and Afghanistan. It's impossible that the CIA thought he was going there as a tourist. If the CIA hadn't caught on to him, it should be dissolved, and its budget used for something worthwhile."[52]

For an example of a sleeper agent within the national media, let's turn to the case of Tarik Hamdi, an Iraqi immigrant who settled in Florida, studied at the University of Tampa, and became affiliated with the Palestinian Islamic Jihad. Hamdi later moved to Herndon, Virginia, where he and his wife, Wafa Huzain, purportedly became agents of HAMAS and al Qaeda.

In 1998 ABC News decided that an interview with Osama bin Laden, who had declared war on America, would be a great way to boost ratings for *Nightline* and granted this assignment to John Miller, one of the network's ace correspondents. Miller required an intermediary to make the necessary arrangements. And so, ABC turned to Tarik Hamdi, who was known for his ties to radical Islam.[53]

Bin Laden, to the network's astonishment, graciously consented to the interview, leading ABC executives to believe that the emir must be either a fan of *Nightline* or else a Muslim leader in search of publicity. No one assumed that the agreement might be based on the fact that bin Laden needed military supplies for his *jihad,* including a replacement battery for his satellite phone.

An order was placed for the battery on May 11, 1998, and the package was sent to Tarik Hamdi's address at 933 Park Avenue in Herndon, Virginia. Two weeks later, Hamdi, with his special package for bin Laden, boarded the chartered plane as a member of the ABC News crew.[54]

Upon his arrival in Pakistan, Hamdi sent the following fax to Khaled al-Fawwaz, an aide to bin Laden: "Brother Khaled: Peace be unto you. We arrived safely and now we are in the Marriott Hotel."[55]

The interview between Miller and bin Laden took place in Afghanistan on May 28, 1998. Osama was all smiles as Miller obsequiously hailed him as a "hero to his followers" and "the Middle East version of Teddy Roosevelt."[56] The emir, in all likelihood, was not only pleased by the praise but also by the supplies that had been delivered to his secret lair by an emissary of ABC.

John Miller's interview with bin Laden was aired on June 8, 1998. On August 7, 1998, the US embassies in Kenya and Tanzania were bombed. Bin Laden would have been incapable of orchestrating this event without Tarik Hamdi and the replacement battery.[57]

Finally, for an example of a "sympathizer to terrorism and other religious fanatics"[58] within the FBI, we need only examine two cases involving special agent Gamal Abdel-Hafiz.

Abdel-Hafiz, like Ali Mohamed, was born in Egypt and spent several years in the Egyptian military before moving to New York. He obtained his US citizenship in 1990, applied for a position with the FBI, and became appointed as special agent and a member of the International Terrorism Squad.[59]

The first troublesome case involving special agent Abdel-Hafiz concerned BMI (Bait-al-Mal, Inc.), a now defunct Islamic bank in Secaucus, New Jersey. Founded in 1983, BMI was financed by known terrorists and members of bin Laden's family. In 1999 an accountant at BMI contacted the FBI to relay his suspicions that millions of the bank's funds had been used to finance al Qaeda's bombings of the US embassies in East Africa.[60] Mark Flessner, the assistant US attorney, was assigned to spearhead the investigation, which bore the bureau code name: Vulgar Betrayal. Flessner requested the assistance of FBI special agent Gamal Abdel-Hafiz. Abdel-Hafiz, after all, was a Muslim, who spoke fluent Arabic, and was a member of the International Terrorism Squad. Since the BMI officials, by and large, were Arabic Muslims, Abdel-Hafiz represented someone who could get to the heart of the matter better than any other agent within the Justice Department. Moreover, Soliman Biheiri, the president of the bank, expressed his willingness to speak to Abdel-Hafiz, since the special agent was sensitive to Muslim issues. Yet when Flessner approached the special agent and asked him to meet with the BMI president and to record the conversation, Abdel-Hafiz expressed outrage and said: "A Muslim does not record another Muslim."[61] Pressed on the matter, Abdel-Hafiz blurted out: "I do not record another Muslim. That is against my religion."[62] When questioned still further about his refusal, Abdel-Hafiz cried out: "Why? I fear for my life!" Reminded that he had the protection of the bureau, Abdel-Hafiz said: "The FBI can't protect me. The FBI, I don't trust them."[63]

Surely these were strange words to come out of the mouth of an FBI special agent. Attorney Flessner, who went on to prosecute the case against BMI, later remarked that Abdel-Hafiz's behavior was highly suspicious. "It's hard to emphasize how odd it was for an FBI agent to refuse to cooperate with an investigation when he had been approached by a grand jury suspect," Flessner said. "It was surreal.

I've never heard it happening in the history of the FBI."[64] John Vincent, an FBI agent who worked with Abdel-Hafiz, added: "He [Abdel-Hafiz] wouldn't have any problems interviewing or recording anybody who wasn't a Muslim, but he couldn't record a Muslim."[65]

When ABC News got the story and inquired about Abdel-Hafiz's refusal to record the statements of a fellow Muslim, the FBI attempted to exonerate the special agent by saying that the clandestine recording of the bank executive would have taken place in a mosque. This was a blatant lie (there was no mosque involved), a fact that the bureau was later forced to admit.[66]

National uproar over the BMI case and the strange behavior of the Muslim special agent brought to light another troublesome case involving Abdel-Hafiz—this one concerned Sami al-Arian, a University of Southern Florida professor. Sami al-Arian became the subject of an FBI probe in 1998 when evidence was unearthed that he was the organizer of a cell of the Palestinian Islamic Jihad in Florida and that he was responsible for transferring vast sums to the terrorist group. In its manifesto, the Palestinian Islamic Jihad (PIJ) referred to the United States as "the Great Satan America" and stated that its objective was to destroy Israel and put an "end to all Western influence" in the Middle East.[67] The terrorist group was responsible for the murder of more than one hundred people in Israel and the occupied territories, including at least two Americans: Alisa Flatow, age twenty, and Shoshana Ben-Yishai, age sixteen.[68]

Knowing that he was the subject of an investigation, Professor al-Arian approached Abdel-Hafiz at a conference in Washington, DC, and questioned him about details of the case. When he returned to FBI headquarters, Abdel-Hafiz informed his colleagues of his meeting with al-Arian. At that time, Barry Carmody, a veteran of the bureau for thirty-four years, asked Abdel-Hafiz to follow up the encounter with a secretly taped telephone conversation. Abdel-Hafiz adamantly refused, saying that he would not record the conversation without Professor al-Arian's knowledge and permission. "That's outrageous," Carmody said. "That defeats the whole purpose."[69]

The case against al-Arian dragged on for years—thanks,

according to Special Agent Carmody, to Abdul-Hafiz's refusal to provide assistance.[70] Finally, on February 20, 2003, the university professor, with ties to radical Islam, was indicted on fifty counts of international terrorism, including money laundering, racketeering, perjury, obstruction of justice, and travel-act violations.[71] Al-Arian was placed without bail in the Coleman Correctional Facility in Sumter County, Florida, where he awaits trial.[72]

What happened to Abdel-Hafiz? How did the FBI respond to the special agent's refusal to tape the conversations of fellow Muslims? Was he drummed out of the bureau for dereliction of duty? Did he become a subject of investigation? Quite to the contrary, Abdel-Hafiz was promoted to the post of deputy legal attaché in Riyadh, Saudi Arabia.[73]

His story, however, lacks the ending of a fractured fairy tale. At the FBI post in Riyadh, Wilfred Rattigan, Abdel-Hafiz's boss, converted to Islam, and the two agents flew off to Mecca for the *hajj*.[74] Complaints soon were raised regarding the failure of the Riyadh office to pursue counterterrorism leads.[75] Bureau officials were dispatched to conduct an audit. They discovered that Abdel-Hafiz had ordered the shredding of two thousand documents relating to 9/11 terror investigations. Several of these documents were not duplicated in FBI computer files, including letters from Saudi security officials about terror suspects.[76] Rattigan and Abdel-Hafiz were recalled to the United States.

Troubles began to mount for Abdel-Hafiz. Bertie Abdel-Hafiz, the agent's former wife, charged that he had faked a burglary at their home in Roanoke, Texas, in order to collect $25,000 in insurance benefits.[77] Abdel-Hafiz was placed on administrative leave, pending a disciplinary review. During the course of the investigation, he failed a lie-detector test when he denied the charges.[78] Despite this, the FBI, in an extraordinary move, opted to overturn the ruling of its senior disciplinary officer and reinstate the agent. Abdel-Hafiz, when contacted by *Newsweek*, said that he was "thrilled" to have his job back and grateful to "have this injustice lifted."[79]

Radical Islam has penetrated nooks and crannies throughout the United States. There are many hardworking and decent Muslim individuals and families who are dedicated Americans and who want to practice their faith in peace and harmony. However, the five thousand or more sleeper agents, who are now in place to plan the American Hiroshima, could be white-collar workers, married to non-Muslims, and living in expensive homes in the suburbs; they could be blue-collar workers, married to fellow Muslims, and living in inner-city apartment complexes; or they could be farm workers, single, and residing in rural settings. They could be attending Christian worship services on Sunday mornings and working at schools, hospitals, chemical plants, nuclear energy facilities, packing houses, industrial warehouses, or government offices. Some, like Ali Abdelsoud Mohamed, might be serving in the armed forces. Others, like Tarik Hamdi, might be working for national news networks. A few might be employed by federal law enforcement agencies, including the FBI. The sleeper agents could be living exemplary lives, according to Islamic precepts, by abstaining from premarital sex, adultery, liquor, drugs, and tobacco. Or else they might be individuals with a weakness for alcohol and tobacco and a taste for pornography and prostitutes. The al Qaeda terrorist who attempted to fly a plane into the fifty-five-story Rialto Towers, Australia's highest building, was a frequent customer at the Main Course brothel in downtown Melbourne, where he became notorious "as a bit of a sneak, always trying to get more than he paid for."[80]

Despite President George W. Bush's statement that the sleepers are "set to go off without warning," the federal government has demonstrated little success in ferreting out the terrorist cells and terrorist agents, even when such cells and agents have been isolated. This failure to adopt appropriate action to protect the American people may be a result of bureaucratic complacency or blatant ignorance of al Qaeda's and radical Islam's avowed intent to destroy the United States of America and to murder millions of Americans for the sake of parity.

CHAPTER TWELVE

Amen, America

The crusaders and the Jews do not understand but the language of killing and blood. They do not become convinced unless they see coffins returning to them, their interests being destroyed, their towers being torched, and their economy collapsing. O Muslims, take matters firmly against the embassies of America, England, Australia, and Norway and their interests, companies, and employees! Burn the ground under their feet, as they should not enjoy your protection, safety, or security. Expel those criminals out of your countries. Do not allow the Americans, the British, the Australians, the Norwegians, and the other crusaders who killed your brothers in Iraq to live in your countries, enjoy their resources, and wreak havoc in them. Learn from your nineteen brothers who attacked America in its planes in New York and Washington and caused it a tribulation that it never witnessed before and is still suffering from its injuries until today. O Iraqi people, we defeated those crusaders several times before and expelled them out of our countries and holy shrines. You should know that you are not alone in this battle. Your mujahid brothers are tracking your enemies and lying in wait for them. The mujahadeen in Palestine, Afghanistan, and Chechnya and even in the heart of America and the West are causing death to those crusaders. The coming days will bring to you the news that will heal your breasts, God willing.

—Ayman al-Zawahiri, May 21, 2003

One of the great failures of the American intelligence community, and especially the counterterrorism community, is to assume if someone hasn't attacked us, it's because he can't or because we've defeated him. Bin Laden has consistently shown himself to be immune from outside pressure. When he wants to do something, he does it on his own schedule.

—Michael Scheuer, former agent in charge of the bin Laden file at the Central Intelligence Agency, November 14, 2004

Bin Laden asserts that he must kill four million Americans—two million of whom must be children—in order to achieve parity for a litany of "wrongs" committed against the Muslim people by the United States of America. The "wrongs" include the establishment and occupation of military bases between the holy cities of Mecca and Medina in Saudi Arabia, the support of Israel and the suppression of the Palestinian people, the Persian Gulf War and the subsequent economic sanctions, and the invasions of Somalia, Afghanistan, and Iraq.

Few military and intelligence officials question bin Laden's ability to carry out this threat. US, Saudi, Pakistani, Russian, Israeli, and British intelligence sources have confirmed that al Qaeda possesses a small arsenal of tactical nuclear weapons—weapons that are being prepared for the "American Hiroshima." Gen. Eugene Habiger, the former executive chief of US Strategic Weapons at the Pentagon, maintains that an event of nuclear megaterrorism on US soil is "not a matter of if, but when."[1]

During the 2004 presidential debates, President Bush and Senator Kerry said that nuclear weapons in the hands of terrorists represents the greatest danger facing the American people, while Vice President Cheney, on the campaign trail, warned that a nuclear attack by al Qaeda appears to be imminent.[2] Moreover, upon leaving office, Attorney General John Ashcroft and Homeland Security Director Tom Ridge both voiced their belief that al Qaeda's plans for an American Hiroshima soon might come to fruition.[3] Strange to say, Ashcroft's statement appeared to contradict his assertion in his letter of resignation to President Bush that "the objective of securing the safety of America from crime and terror has been achieved."[4]

From the private sector, billionaire Warren Buffett, who establishes odds against cataclysmic events for major insurance companies, concluded that an imminent nuclear nightmare within the United States is "virtually a certainty."[5]

From the academic community, Dr. Graham Allison, the director of Harvard University's Belfer Center for Science and International Affairs, warns: "Is nuclear megaterrorism inevitable? Harvard professors are known for being subtle or ambiguous, but I'll try to be clear. 'Is the worst yet to come?' My answer: Bet on it. Yes."[6]

Such warnings, by and large, have been ignored by the national media. This is particularly surprising in the wake of a think piece on the seriousness of the threat by Bill Keller, editor of the *New York Times*. Keller, as noted earlier, wrote that the only reason for anyone to think that the nuclear attack won't happen is because "it hasn't happened yet"—adding such reasoning represents "terrible logic."[7]

At present, the consensus of global intelligence about the nuclear capacity of al Qaeda can be summarized as follows:

1. There are fully assembled nuclear weapons in bin Laden's arsenal. The only disagreement comes with the number. The Russians say twelve to fifteen; the Saudis claim forty to seventy.
2. Bin Laden obtained these weapons through his connections with the Chechen rebels and the Russian Mafia.
3. The Chechen rebels helped to recruit Soviet scientists and SPETSNAZ technicians so that the weapons could be properly assembled and maintained.
4. The location of the weapons remains unknown, but a stockpile was in Afghanistan before the launching of Operation Enduring Freedom on October 7, 2001.
5. Several nuclear weapons, including suitcase bombs, mines, rucksacks, and crude tactical nuclear warheads, have been forward-deployed to the United States.
6. Thousands of sleeper agents are estimated to be in place throughout the United States.

7. Many of the agents, including Adnan el-Shukrijumah, Anas al-Liby, Jaber A. Elbaneh and Amer el-Maati, have been trained in nuclear technology.
8. Nuclear supplies and materials have been transported across the Mexican border into the US.
9. The next attack is planned to occur simultaneously at various sites throughout the country. Certain targets include Boston, New York, Washington, DC, Las Vegas, Miami, Chicago, and Los Angeles.
10. Muslim terrorist organizations throughout the world, including Hezbollah, have been aiding al Qaeda in this massive undertaking.

Any lingering doubts among skeptics about bin Laden's Manhattan Project should have been put to rest by Michael Scheuer, the CIA agent who had been in charge of the bin Laden files (code-named "Alec"). On November 14, 2004, Scheuer appeared on *60 Minutes*, the CBS newsmagazine, alerting the American people that a nuclear attack by al Qaeda "is pretty close to being inevitable."[8]

Scheuer further confirmed that bin Laden had met with Islamic religious leaders and had obtained from Hamid bin Fahd, a Saudi sheikh, a ruling that permitted the use of nuclear weapons against the United States of America. This ruling was granted by Fahd on behalf of other Wahhabist clerics so that the American people will receive just retribution for their numerous offenses against Muslims throughout the world.[9]

What will happen? For bin Laden, who believes in parity, the model is Hiroshima. The very weapon that the United States unleashed against the civilian population of Japan, he believes, should be used to kill millions of Americans. This is in keeping with the following teaching: "Unbelievers are those who do not judge according to God's revelations. We decreed for them a life for a life, an eye for an eye, a nose for a nose, an ear for an ear, a tooth for a tooth, and a wound for a wound." The atomic bomb that was dropped on August 6, 1945, at 8:15 AM had an explosive yield of twelve kiloton. The conventional explosives wiped out several blocks upon impact and gave rise to a massive fireball, the core of which reached a maximum temperature of ten million degrees Celsius (18,003,201 degrees Fahrenheit). As it expanded from ground zero, the temperature of the fireball dropped to that of the temperature of the sun. Within a radius of a half mile, everything—people, animals, houses, buildings, cars—vanished in the blink of an eye. Birds ignited and disappeared in midair. Metallic objects within 450 feet of the explosion vaporized. Because light travels faster than sound, the victims were incinerated before they could hear the sound of the explosion. All that remained of them were shadows burned into stone.

The expanding shock waves from the blast sucked oxygen into the fireball, propelling it to encompass more of the city. Thousands within a radius of a mile died of intensive asphyxiation and lung burns within minutes.

The bomb expended 35 percent of its energy in the form of this radiated heat. An additional 50 percent became absorbed into the atmosphere to become a juggernaut blast—a wave that ripped through Hiroshima at 670 miles per hour. The buildings that survived the melting heat became twisted like pretzels by the force of the incredible wind. Thousands more were killed by the deadly hail of debris and shattered glass.

The explosion created a huge mushroom-shaped cloud of irradiated debris that rose two to three miles in the air. Within forty minutes, the debris began to fall back to earth, showering the injured survivors and dooming the rescue workers. Effects of radiation are measured in millirems. A five-hour airplane ride delivers three millirems to every passenger; a chest x-ray delivers ten.[10] All those within a half mile of the blast in Hiroshima were exposed to 10,000,000 millirems and instant death. People three-quarters of a mile away received a dose of 1,000,000 millirems, meaning certain death in days, if not hours. Those farther out—from a mile to a mile and a half from ground zero—were irradiated with 500,000 millirems, killing half within a matter of months. Five percent of those exposed to 100,000 millirems also died within a short period of time. One in a hundred of those exposed to 12,500 millirems developed inoperable malignancies.[11]

Two hundred thousand inhabitants of Hiroshima were dead by the end of 1945. Others were left horribly wounded—blinded by the blast and burned beyond recognition. The city had been razed; multistoried buildings had been reduced to charred posts.

Visiting the city thirty days after the blast, Wilfred Burchett, an Australian journalist, wrote: "Hiroshima does not look like a bombed city. It looks as if a monster steamroller has passed over it and squashed it out of existence." Within the hospitals, Burchett became one of the first witnesses to radiation sickness. He described patients with "purple skin hemorrhages," gangrene, fever, and rapid hair loss. "In Hiroshima," he wrote, "thirty days after the first atomic bomb destroyed the city and shook the world, people are still dying mysteriously and horribly—people who were uninjured in the cataclysm from something which I can only describe as the atomic plague."[12]

Hiroshima, with its population of 350,000, wasn't nearly as densely populated as New York City, with a population of 8 million. If one of bin Laden's ten-kiloton nukes had been detonated on September 11, 2001, the World Trade towers, all of Wall Street and the financial district, along with the lower tip of Manhattan up to Gramercy Park, and much of midtown, including the theater district, would lay in ruins. Tens of thousands would melt like raindrops on hot pavement. The dustlike contamination, under ordinary weather conditions, would ride for five to ten miles on prevailing winds, deep into the Bronx or Queens or New Jersey.[13] Up to 50 percent of the exposed population would die an agonizing death of radiation poisoning within the subsequent days and weeks.

In the wake of such a disaster, New York would become an uninhabitable wasteland. The radioactive buildings and streets would have to be demolished and the contaminated debris and topsoil removed. But irregular radiation and hot spots would persist for hundreds, if not thousands, of years.[14] By conservative estimates, millions would die; millions more would be left permanently blind, covered with third-degree burns, and badly crippled.

Even the effects of one of bin Laden's smallest tactical nukes—a device with an explosive yield of one kiloton—would produce a nightmare almost beyond comprehension. If such a weapon were detonated in Times Square, the blast would collapse buildings for two blocks in every direction; the searing heat would transform pedestrians into mounds of ashes; and the shock wave would crush workers at their desks.[15] The final toll of the dead would exceed 250,000; the number of severely burned and injured over 500,000.

There would be other effects to such an attack. The financial and cultural center of America would cease to exist. The GNP would drop more than 3 percent in a matter of seconds.[16] One of America's major ports would be closed indefinitely. Millions of Americans would lose their jobs. At the same time, the number of wounded and traumatized would tax the country's healthcare systems. Makeshift hospitals would be set up in schools, museums, libraries, and other public buildings. Many victims would languish for days in unspeakable pain and suffering and without basic medical care. Populations would desert major metropolitan areas. Within hours of the blast, the US economy would fall into a deep depression from which it might never recover.

This is an optimistic scenario. It depicts the detonation of one device in one American city. Bin Laden, according to the testimony of Khalid Shaikh Mohammed and other al Qaeda witnesses, has announced his intention to detonate at least seven of these devices at various strategic locations throughout the country.[17] He believes that such an event is necessary to issue forth the Day of Islam—the day when all of creation bows in fear and trembling before the throne of Allah.

The "best-case" scenario, according to Dr. Graham Allison and other experts, is that al Qaeda will only succeed in setting off a radiological dispersion device, commonly known as a "dirty bomb." Such a weapon would be quite easy to assemble. The requisite elements would be radioactive material, a few sticks of dynamite, and a suicide bomber, seeking instant martyrdom and a heavenly reward of seventy beautiful, willing, and virginal *houris* to cater to his every whim for all eternity. The terrorist could import his own radioactive material or choose from a variety of goods available at tens of thousands of places throughout the country, including food-irradiation plants, universities, medical facilities, industrial plants, and x-ray laboratories. To complete his mission, the al Qaeda agent would wrap the radioactive material within the dynamite, pack the device in a suitcase, travel to the designated location (such as the Las Vegas strip, Century City in L.A., or Times Square), open the case, strike a match, and cry out: "Allahu Akbar!" It would be that simple

In 2001 the National Regulatory Commission reported that radioactive material from 835 sites disappeared within the past five years. The missing materials included cobalt-60, used in industrial radiography and radiotherapy in hospitals, and cesium-137, used for the sterilization of food products, including wheat, spices, flour, and potatoes, and as a treatment for cancer. These substances emit deadly gamma rays that could penetrate the skin causing immediate cellular damage and subsequent death by radiation poisoning or various forms of cancer

The Center for Strategic and International Studies, a Washington think tank, came up with a scenario of a bomb made of cesium-137 from a food-irradiation plant and a truck stuffed with two tons of TNT being set off near the National Mall in Washington, D. C.[18] In the best of circumstances, only hundreds would be killed from the conventional explosion. But the blast would carry the radioactive particles hundreds of feet into the air. A prevailing light wind would carry the cesium particles over several miles of the metropolitan area and into Maryland and Virginia.[19] Within minutes, these particles would begin to fall upon people who would be unaware of the danger. The microscopic and hence invisible particles would adhere to everything: skin, clothing, cars, buses, buildings, sidewalks, and the soles of shoes.

A single grain of cesium, if inhaled or ingested, is a bone sucker. It heads to the marrow to stop the production of platelets and to cause the mutation of cells. Within hours or days of the blast, many of those exposed to these particles could suffer the first symptoms of radiation poisoning: bleeding from the nose and mouth, diarrhea, nosebleeds, hair loss, purpled fingernails, high fever, and delirium.[20] Several thousand would be dead within a matter of weeks. Their bodies would have to be buried within lead-lined caskets. For those who would survive, cesium produces mutations in cell formation.[21] Within a matter of a few years, hundreds of thousands would develop various forms of malignancies: thyroid cancer, cancer of the pancreas and liver, breast cancer, cancer of the reproductive organs, and brain tumors.

The attack would produce other results. The decontamination effort would require that the radioactive particles be removed. Such particles cannot be blown or washed away. The soil throughout the city would have to be covered with mounds of sands; trees and shrubbery hacked down; and household pets put to death. Key government buildings and national monuments would have to be razed. Mass evacuation and resettlement would be required. Medical facilities within DC and surrounding states would become crammed with the injured and dying. Congress would go into recess. Many emergency workers, fearful of exposure to the deadly gamma rays, would refuse to return to the city. The mass transit system would be shut down. Vehicles would be confiscated for fear they might carry the deadly particles to other locations. The airports, including Dulles, would close. Real estate prices would plummet. Residences would become abandoned. Places of employment would close never to reopen. Hotels, stores, and restaurants would face the wrecking ball. Unemployment would soar. Martial law would be imposed.[22] The end result, according to Henry Kelly, president of the Federation of American Scientists, would be losses of "trillions of dollars."[23]

"If a dirty bomb attack becomes the best shot that al Qaeda can take," a national security official said, "the people of America should declare victory."[24]

When will it happen? Al Qaeda, according to US military analysts, places a great deal of significance on dates. The bombing of the World Trade Center on February 26, 1993, was planned to coincide with the second anniversary of the launching of the ground war in Operation Desert Storm on February 23, 1991. The date of September 11 was significant since it represented the fifth anniversary of the conviction in a New York court of World Trade Center bomber Ramzi Yousef. The bombings of the US embassies in Nairobi, Kenya, and Dar es Salaam, Tanzania, occurred on August 7, 1998, the eighth anniversary of President George H. W. Bush's 1990 commitment to deploy troops for the protection of Saudi Arabia against Saddam Hussein. And the ill-fated attack on the warship USS *The Sullivans* occurred on January 3, 2000, in celebration of the holiest day of Ramadan.

Some analysts believe that the favored month for the next attack is October. October 2 looms ominous, since it is the anniversary of the federal court conviction of blind Sheikh Omar Abel Rahman, the spiritual mentor of bin Laden and Ayman al-Zawahiri. Another troublesome date is October 7, the day of the launching of Operation Enduring Freedom.

Another potential date would be August 6 to commemorate the day, in 1945, when *Enola Gay*, the crew-christened name of a US B-29 bomber, dropped a twenty thousand-pound atomic bomb called "Little Boy" on a place called Hiroshima.

EPILOGUE

The Doomsday Clock

With nuclear weapons-grade uranium, the background neutron rate is so low that terrorists, if they have such material, would have a good chance of setting off a high-yield explosion simply by dropping one half of the material onto the other half. Most people seem unaware that if separated highly enriched uranium is at hand, it's a trivial job to set off a nuclear explosion . . . even a high school kid could make a bomb in short order. Such a bomb could have a yield of 15 kilotons.
 —Luis W. Alvarez, Manhattan Project physicist

All things come from Allah. The atomic bomb comes from Allah, so it must be used.
 —Student within Pakistan's Haqquania Madrassah to
 New York Times reporter Jeffrey Goldberg

We bow our heads to Allah almighty for restoring greatness to Pakistan on May 28, 1998.
 —Atta-ur-Rahman, Pakistan's minister of science

The majority of Al Qaeda's active leadership in the land of two temples— the mujahadeen—need not expand its religious consciousness or make further preparations because it has already been trained to be prepared.

Praise Allah, the message of the sheikh has reached the general leader-ship of the Muslims and mujahadeen throughout the world and those orders will be understood only by those who are part of the mujahadeen. And wait, salvation is near and the painful blow is on the way. It will be sweeping, and with God's help and signs recognized by those who believe in their hearts in jihad and the mujahadeen. And God will proclaim par-adise and our sheikh will take the Islamic army to his bosom and its flag shall fly everywhere in the world. We are waiting.

—Abu al Bara'a al-Qarshy, one of Osama bin Laden's most
trusted lieutenants, February 12, 2003

The Doomsday Clock was designed by scientists from the Man-hattan Project to serve as a barometer of nuclear danger and as a call for nuclear disarmament. Throughout the years, this actual physical object, which represents the state of nuclear danger, has been updated and maintained by the *Bulletin of Atomic Scientists* in Wash-ington, DC. The members of this body decide when to move the clock hands and how far to move them in symbolizing the world's vulnerability to nuclear danger. The clock started figuratively to tick in 1947, the beginning of the cold war, when the Soviet Union ini-tiated its quest to become a nuclear superpower.

The minutes of the clock began to inch toward midnight, the hour of nuclear holocaust, as the Soviet Union conducted its first successful nuclear explosion at its Semipalatinsk site in Kazakhstan on August 29, 1949. The minute hand moved forward as Great Britain exploded a nuclear device at Monte Bello Islands near Aus-tralia on October 3, 1952, and as the United States exploded the first H-bomb on November 1, 1952. It moved further when France became a member of the nuclear club by detonating an atomic bomb in the Sahara Desert on February 13, 1960.

At first, the American people were alarmed by the ticking of the clock. Some built bomb shelters in their basements to protect their families from the nuclear blast. The shelters were equipped with food, water, and shortwave radios so they could know when it would be safe to emerge into the postapocalyptic world. Schools conducted drills so that children would know how to crawl under their desks in fetal positions to protect themselves from falling rubble. Hospitals and pharmacies remained stocked up with potassium iodine tablets to protect against radiation poisoning. Books and movies such as *Fail Safe* and *On the Beach* alerted the general public to the gravity of the threat.

The fear of the coming Doomsday was exacerbated by the Cuban missile crisis of October 16–28, 1962; Red China's successful nuclear explosion test on October 16, 1964; and Israel's secret development of nuclear weapons, thanks to help from the French, at Dimona in the Neyev desert from 1960 to 1967. More books and movies were released, some serious and some satirical, including *The Bedford Incident* and *Dr. Strangelove, or, How I Learned to Stop Worrying and Love the Bomb*, to underscore the theme of humanity's vulnerability to a nuclear catastrophe. Two of the most popular songs of 1965—"Eve of Destruction" and "Just a Little Rain"—drove home the same message.

On July 1, 1968, the scientists, overseeing the Doomsday Clock, slowed the minute hand, because of the signing of the Nuclear Nonproliferation Treaty. This treaty prohibited nonnuclear states from building or acquiring atomic weapons and obligated the various nuclear powers (the United States, the Soviet Union, Great Britain, and France) to work on arms control and disarmament. The threat no longer seemed imminent.

But such hope was illusionary. On May 18, 1974, India conducted a nuclear explosion in the Rajasthan Desert, claiming it was for peaceful purposes. In the wake of this development, new efforts were enacted to ward off the threat of nuclear proliferation and the development of more powerful nuclear weapons. The Peaceful Nuclear Explosion Treaty, which prohibited nuclear explosions with yields more than 150 kilotons, was signed on May 28, 1976.

Though the cold war was coming to an end, there were other flair ups of concern when the clock could be heard ticking. One such event took place on July 20, 1982, when President Reagan's hardline stance toward the Soviet Union caused him to withdraw the United States from negotiations for a comprehensive test-ban treaty.

Another incident occurred on February 16, 1987, when the Soviets, after a nineteen-month moratorium, proceeded to conduct new tests of new nuclear weapons, including the Tsar Bomba ("King of Bombs"), a 100-megaton weapon. But no one really wanted to play the nuclear hand in war or combat—not even with tactical nukes. MAD, or Mutually Assured Destruction, seemed to keep everyone in check, since firing a nuclear missile meant getting one or more back in return—in essence, suicide. Even during the Soviet conflict in Afghanistan, no nuclear devices were used. After this conflict came to an end in 1989, the nuclear club of nations (the United States, the Soviet Union, China, Great Britain, France, India, and Israel) was ready for *perestroika*.

Something amazing happened. The clock stopped at fifteen minutes to midnight. The hands appeared to have been frozen by the calls of the United Nation for the Comprehensive Test Ban Treaty, which was ratified on September 24, 1996.

The directors of the *Bulletin of Atomic Scientists* were shook from their slumber by the five nuclear weapons that were tested by Pakistan beneath the scorched hills of the Baluchistan Desert on May 28, 1998. The clock was now reset at nine minutes to midnight.

On February 27, 2002, the hand moved two minutes closer to midnight in the wake of growing concerns about the security of nuclear materials throughout the world and reports that al Qaeda might have acquired these materials to produce nuclear weapons. "The confluence of the rise of extremists who sacrifice their lives for their cause combined with weapons of mass destruction is an especially worrisome development," the directors said.[1]

Since the time of this announcement, the minute hand should have been pressing forward toward the Doomsday hour. Its reaching toward midnight should have been hastened by other events: (1) the news that Dr. A. Q. Khan has provided nuclear technology and assistance to an unsavory host of countries (North Korea, Iran, Libya, Saudi Arabia, Iraq, Sudan, Nigeria, Brazil, Egypt, Malaysia, Indonesia, Algeria, Kuwait, Myanmar, and Abu Dhabi) and to al Qaeda; (2) the statements of intelligence officials, including Michael Scheuer, the CIA agent in charge of the bin Laden file, that al Qaeda has bought and developed nuclear weapons; (3) the arrest of a terrorist with a nuclear device in Ramallah; (4) the information derived from the interrogations of Khalid Shaikh Mohammed and scientists from the A. Q. Khan Research Laboratories in Pakistan about bin Laden's plans for an American Hiroshima; and (5) public pronouncements by al Qaeda officials, including bin Laden and Al-Zawahiri, that the terror group possesses an arsenal of tactical nukes that will be used in the great *jihad* against the United States of America.

When will the clock strike midnight? No one knows for sure. It could occur within a month or a year or two. But most experts believe that it will happen soon. As Bill Keller of the *New York Times* came to conclude in his think-piece "Nuclear Nightmares," the only reason for thinking that a nuclear and/or radiological attack won't happen is simply because "it hasn't happened yet."[2] As this book goes to press, millions of Americans may be living on borrowed time.

APPENDIX

Al Qaeda's Search for Weapons
of Mass Destruction

AL QAEDA'S WMD ACTIVITIES

Time of Incident	WMD	Incident	Sources
Unspecified	Nuclear	A leaked intelligence report states that bin Laden paid over two million British Pounds to a middleman in Kazakstan for a "suitcase" bomb.	Marie Colvin, "Holy War with US in His Sights," *Times* (London), August 16, 1998.
Unspecified	Nuclear	Bin Laden gave a group of Chechens $30 million in cash and two tons of opium in exchange for approximately twenty nuclear warheads.	Riyad 'Alam al-Din, "Report Links Bin Laden, Nuclear Weapons," *Al-Watan al-Arabi*, November 23, 1998; Emil Torabi, "Bin Laden's Nuclear Weapons," *Muslim Magazine* (Winter 1998).

Time of Incident	WMD	Incident	Sources
Unspecified	Biological	Associates of bin Laden bought anthrax and plague from arms dealers in Kazakstan.	Paul Daley, "Report Says UBL-Linked Terrorist Groups Possess 'Deadly' Anthrax, Plague Viruses," *Melbourne Age*, June 4, 2000.
Unspecified	Nuclear/ Radiological	Bin Laden sent envoys to several eastern European countries to purchase enriched uranium.	"Arab Security Sources Speak of a New Scenario for Afghanistan: Secret Roaming Networks That Exchange Nuclear Weapons for Drugs," *Al-Sharq al-Awsat*, December 24, 2000.
Unspecified	Nuclear/ Radiological	Bin Laden obtained seven enriched-uranium rods from Ukrainian arms dealer Semion Mogilevich.	Uthman Tizghart, "Does Bin Laden Really Possess Weapons of Mass Destruction? Tale of Russian Mafia Boss Simion Mogilevich Who Supplied Bin Ladin with the Nuclear 'Dirty Bomb,'" *Al-Majallah* (London), November 25, 2001.

Unspecified	Biological	US officials discover documents concerning the aerial dispersal of anthrax via balloon within the Kabul office of Pakistani scientist Dr. Bashiruddin Mahmood.	"Sketches of Anthrax Bomb Found in Pakistani Scientist's Office," *Rediff.com*, November 28, 2001.
Unspecified	Chemical/ Biological	Al-Qaeda's five-thousand-page *Encyclopedia of Jihad* is devoted to construction of CBW (chemical and biological weapons).	"Osama Bin Laden's Bid to Acquire Weapons of Mass Destruction Represents the Greatest Threat That Western Civilization Has Faced," *Mail on Sunday* (London), June 23, 2002.
Unspecified	Chemical	CNN correspondent Mike Boettcher reports that US intelligence agencies have discovered evidence of recent purchases of cyanide by al Qaeda operatives.	"Wolf Blitzer Reports," CNN, July 31, 2002.

Time of Incident	WMD	Incident	Sources
Unspecified	Nuclear/ Biological/ Chemical	Two Pakistani scientists shared nuclear, biological, and chemical weapons information with bin Laden and learned of nuclear material that had been provided to him by the Islamic Movement of Uzbekistan.	Toby Harnden, "Rogue Scientists Gave Bin Laden Nuclear Secrets," *Daily Telegraph* (London), December 13, 2001; Peter Baker, "Pakistani Scientist Who Met Bin Laden Failed Polygraphs, Renewing Suspicions," *Washington Post*, March 3 2002; Susan B. Glasser and Kamra Khan, "Pakistan Continues Probe of Nuclear Scientists," *Washington Post*, November 14, 2001.
1993–1994	Nuclear/ Radiological	Jamal al-Fadl claims that, on behalf of bin Laden, he arranged the purchase of uranium for nuclear weapons.	Kimberly McCloud and Matthew Osborne, "WMD Terrorism and Usama Bin Laden," CNS Report, November 20 2001.
1996–1998	Chemical	Bin Laden purchased CW (chemical	Muhammad Salah, "Bin Ladin Front Reportedly Bought

		weapons) over a two-year period prior to 1998 from European states and the former Soviet Union. This information comes from the testimony of a *jihad* leader arrested on August 20, 1998, in Baku, Azerbaijan.	CBW from E. Europe," *Al-Hayah*, April 20, 1999; Muhammad Salah, "US Said Interrogating Jihadist over CBW,"*Al-Hayah*, April 21, 1999.
1997–1998	Chemical/ Biological	Islamic extremists, including al Qaeda operatives, are trained in secret camps near Baghdad in how to use CW and BW (biological weapons) by instructors from the secret Iraqi military intelligence Unit 999.	Gwynne Roberts, "Militia Defector Claims Baghdad Trained Al-Qaeda Fighters in Chemical Warfare," *Sunday Times* (London), July 14, 2002.
October 1997	Chemical/ Biological	A meeting is held in Sudan	Jihad Salim, "Report on Bin

Time of Incident	WMD	Incident (continued)	Sources (continued)
		between bin Laden, Ayman al-Zawahiri, and Hasan al-Turabi, leader of Sudan's National Islamic Front regime, about the construction of a CBW factory.	Ladin, Zawahiri, 'Afghans,'" *Al-Watan al-Arabi*, February 16, 2001.
1998	Nuclear/ Radiological	Russian intelligence reportedly blocked a deal by bin Laden to purchase Soviet-origin uranium.	Earl Lane and Knut Royce, "Nuclear Aspirations? Sources: Bin Laden Tried to Obtain Enriched Uranium," *Newsday*, September 19, 2001.
1998	Chemical/ Biological	A reporter obtains two computers from looters in Kabul that had been found in an abandoned al Qaeda safe house. One of	Alan Culluson and Andrew Higgins, "Computer in Kabul Holds Chilling Memos," *Wall Street Journal*, December 31, 2001; "Report: Al Qaeda Computer Had Plans for

the computers contains a file describing "plans to launch a chemical and biological weapons program." Bin Laden's deputy al-Zawahiri reportedly created computer documents describing his CW and BW program, codenamed "Curdled Milk." The document includes work on a pesticide/ nerve agent that was tested on rabbits and dogs. Al-Zawhahiri was assisted by Midhat Mursi (aka Abu Khabbab), a chemical engineer.

Bio-Weapons," Reuters, December 21, 2001.

May 1998	Chemical/ Biological	Al Qaeda purchases three CBW factories in the former Yugoslavia and	Guido Olimpio, "Islamic Group Said Preparing Chemical Warfare on the West,"

Time of Incident	WMD	Incident (continued)	Sources (continued)
		hires a number of Ukrainian chemists and biologists to train its members.	*Corriere della Sera,* July 8, 1998; Yossef Bodansky, *Bin Laden: The Man Who Declared War on America* (New York: Prima, 2001), p. 326.
August 1998	Chemical	The CIA discovers that bin Laden had attempted to acquire unspecified CW for use against US troops stationed in the Persian Gulf.	Barry Schweid, "US Suggests Iraq Got Weapons from Sudan," *Record* (New Jersey), August 27, 1998.
September 1998	Nuclear/ Radiological	Mamdouh Mahmud Salim, an al Qaeda operative, is arrested in Munich, Germany, for trying to purchase nuclear material, including highly enriched uranium.	Benjamin Weiser, "US Says Bin Ladin Aide Tried to Get Nuclear Weapons," *New York Times,* September 26, 1998.

September 1998	Chemical	Wadi al-Hajj, a Lebanese national, is arrested in Arlington, Texas, for perjury. The FBI contends that he had lied about his affiliation with bin Laden and that he was involved in procuring WMD for al Qaeda.	CNN, December 20, 1998.
December 1998	Chemical/ Nuclear	In an interview with *Time* magazine, bin Laden says that acquiring weapons of any type, including chemical and nuclear, is a Muslim "religious duty."	"Interview with bin Laden," *Time*, December 24, 1998.
1999	Chemical	Afghan sources maintain that bin Laden is using a plant in Charassiab, a district south of Kabul, to produce CW.	"Afghan Alliance— UBL Trying to Make Chemical Weapons," *Parwan Payam-e Mojahed*, December 23, 1999.

Time of Incident	WMD	Incident	Sources
April 1999	Biological	Bin Laden obtains BW substances through the mail from countries of the former Soviet Union (the Ebola virus and salmonella bacterium), from East Asia (anthrax-causing bacteria), and from the Czech Republic (botulinum toxin).	Al J. Venter, "Elements Loyal to Bin Laden Acquire Biological Agents 'Through the Mail,'" *Jane's Intelligence Review* (August 1999); Khalid Sharaf al-Din, "Bin Ladin Men Reportedly Possess Biological Weapons," *Al-Sharq al-Awsat*, March 6, 1999.
June 1999	Chemical/ Biological	Bin Laden has constructed "crude" CBW laboratories in Khost and Jalalabad, Afghanistan, and has acquired the necessary ingredients for CW and BW from former Soviet states.	John McWethy, "Bin Laden Set to Strike Again?" ABC News, June 16, 1999.

July 1999	Chemical/ Biological	An Islamist lawyer states that al Qaeda has CBW and will likely use such weapons against the United States.	"Islamist Lawyer on Bin Laden, Groups," *Al-Sharq al-Awsat*, July 12, 1999.
February 2000	Chemical	A plot by nine Moroccans, with ties to al Qaeda, to poison the water supply of the US Embassy in Rome using a cyanide compound is foiled by Italian police.	Eric Croddy et al., "Chemical Terrorist Plot in Rome?" *CNS Research Story*, March 11, 2002.
February 2, 2000	Chemical/ Biological	CIA director George Tenet informs the Senate that bin Laden has displayed a strong interest in CW and that his operatives have been "trained to conduct attacks with toxic chemicals or biological toxins."	Pamela Hess, "Al Qaeda May Have Chemical Weapons," United Press International, August 19, 2002.

Time of Incident	WMD	Incident	Sources
Late 2000	Nuclear	The intelligence agency of an unnamed European country intercepts a shipment of approximately twenty nuclear warheads—originating from Kazakstan, Russia, Turkmenistan, and the Ukraine. The shipment was intended for bin Laden and the Taliban regime of Afghanistan.	"Arab Security Sources Speak of a New Scenario for Afghanistan: Secret Roaming Networks That Exchange Nuclear Weapons for Drugs," *Al-Sharq al-Awsat*, December 24, 2000.
2001	Biological	Various reports maintain that Muhammad Atta, the leader of the September 11 hijackers, was provided with a vial of anthrax by an Iraqi intelligence agent at a meeting in Prague.	Kriendler and Kreindler 9/11 lawsuit; "Prague Discounts an Iraqi Meeting," *New York Times*, October 21, 2001; "Czechs Retract Iraq Terror Link," United Press International October 20, 2001.

2001	Biological	Ahmed Ressam, arrested in a plot to bomb LAX, testifies that bin Laden is interested in using low-flying aircraft to dispense BW agents over major metropolitan areas of the United States.	"Bin Laden's Biological Threat," BBC, October 28, 2001.
2001	Biological	Documents found in Afghanistan reveal that al Qaeda was conducting research on using botulinum toxin to kill two thousand people.	"Al Qaeda Tested Germ Weapons," Reuters, January 1, 2002.
2001	Chemical	Ahmed Ressam says that he had witnessed the gassing of a dog with cyanide in an al Qaeda training camp.	Pamela Hess, "Al Qaida May Have Chemical Weapons," United Press International, August 19, 2002.

Time of Incident	WMD	Incident	Sources
February 2001	Chemical	The United States aborts a planned air strike against Afghanistan for fear of a chemical attack by al Qaeda, after receiving warnings from an Arab embassy in Islamabad, Pakistan.	Sa'id al-Qaysi, "US Said Aborted Planned Attack on Bin Laden for Fear of 'Chemical Strike,'" Al-Watan al-Arabi, February 16, 2001.
February 2001	Chemical	Bin Laden's elite 055 Brigade is reorganized under the leadership of Midhat Mursi, aka Abu Khabab, an Egyptian and expert in sarin gas production.	Sa'id al-Qaysi, "US Said Aborted Planned Attack on Bin Ladin for Fear of 'Chemical Strike'," Al-Watan al-Arabi, February 16, 2001.
April 2001	Nuclear/ Radiological	Ivan Ivanov claims he met bin Ladin in China to discuss the establishment of a company to buy nuclear	Adam Nathan and David Leppard, "Al-Qaeda's Men Held Secret Meetings to Build 'Dirty Bomb,'" Sunday Times (London), October 14, 2001.

		waste. Ivanov was then approached by a Pakistani chemical engineer interested in buying nuclear fuel rods from the Bulgarian Kozlodui reactor.	
Since summer 2001	Chemical/ Biological/ Nuclear	Iraqi military instructors provided training to al Qaeda operatives in northern Iraq in the use of CBW and the handling of nuclear devices. Between 150 and 250 al Qaeda agents received instruction.	"Abu Nidal's Nemesis," DEBKA file (Jerusalem), August 20, 2002.
Before September 11, 2001	Nuclear	Bin Laden purchases forty-eight "suitcase nukes" from the Russian Mafia.	"Al-Majallah Obtains Serious Information on Al-Qaeda's Attempt to Acquire Nuclear Arms," *Al-Majallah* (London-based Saudi weekly), September 8, 2002.

Time of Incident	WMD	Incident	Sources
October 2001	Nuclear	Mossad arrests al Qaeda operative with backpack purportedly to be a tactical nuclear weapon at the checkpoint in Ramallah.	United Press International, December 21, 2001. First reports spoke of a radiological bomb.
Before November 2001	Chemical	CNN releases videotapes, made by al Qaeda that show dogs being killed by unidentified toxic chemicals that experts believe could be either a crude nerve agent or hydrogen cyanide gas.	"Insight," CNN, August 19, 2002.
November 2001	Chemical/ Nuclear	In an interview bin Laden says: "We have chemical and nuclear weapons as a deterrent and if America used them	Hamid Mir, "Osama Claims He Has Nukes: If US Uses N-Arms It Will Get Same Response," *Dawn* (Pakistan), November 10, 2001.

November 2001	Nuclear	against us we reserve the right to use them."	
November 2001	Nuclear	Evidence gleaned from the offices of Ummah Tameer E-Nau in Kabul shows that a nuclear weapon may have been shipped to the United States from Karachi in a cargo container.	Arnaud de Borchgrave, "Al Qaeda's Nuclear Agenda Verified," *Washington Times*, December 10, 2001.
November 2001	Nuclear	Bin Laden acquires a Russian-made suitcase nuclear weapon from central Asian sources. The weapon is said to weigh 8 kg and to possess at least 2 kg of fissionable uranium and plutonium. The report said the device, with a serial number of 9999 and a manufacturing	"N-weapons May Be in US Already," *Daily Telegraph* (Sydney, Australia), November 14, 2001.

Time of Incident	WMD	Incident (continued)	Sources (continued)
		date of October 1998, could be set off by a mobile phone signal. This weapon, sources say, has been forward-deployed to the United States.	
November 2001	Nuclear	A *Times* (London) reporter discovers a blueprint for a "Nagasaki bomb" within an abandoned al Qaeda house in Kabul.	"Nuke Plans Found; Brit Paper Discovers Details of Weapons in Kabul Safe House," *Toronto Sun,* November 15, 2001; Hugh Dougherty, "Afghan Nuclear Weapons Papers 'May Be Internet Spoofs,'" Press Association, November 19, 2001.
November 2001	Nuclear	"Superbomb" manual that discusses the advanced physics of nuclear weapons and dirty bombs is discovered	"Osama Bin Laden's Bid to Acquire Weapons of Mass Destruction Represents the Greatest Threat That Western Civilization Has Faced," *Mail on*

		in a safe house in Afghanistan.	*Sunday* (London), June 23, 2002.
December 2001	Radiological	Uranium 235 discovered within a lead-lined canister in Kandahar.	Barbie Dutter and Ben Fenton, "Uranium and Cyanide Found in Drums at Bin Laden's Base," *Daily Telegraph* (London), December 24, 2001.
Late 2001	Biological	US operatives discover evidence in Afghanistan that one or more Russian scientists were helping al Qaeda develop biological weapons, including anthrax.	Jeffrey Bartholet, "Terrorist Sleeper Cells," *Newsweek*, December 9, 2001.
Late 2001	Biological	Reports claim that al-Zawahiri's home in Kabul tested positive for traces of anthrax, as did five of nineteen al Qaeda labs in Afghanistan.	"Al-Qaeda: Anthrax Found in Al-Qaeda Home," Global Security Newswire, December 10, 2001; Judith Miller, "Labs Suggest Qaeda Planned to Build Arms, Officials Say," *New York*

Time of Incident	WMD	Incident	Sources (continued)
			Times, September 14, 2002.
Late 2001	Biological	John Walker Lindh told interrogators that a biological attack was expected to be a "second wave" al Qaeda attack.	"Walker Lindh: Al Qaeda Planned More Attacks," CNN, October 3, 2002.
2002	Chemical	The facility of Ansar al-Islam, a radical Islamist group operating in northern Iraq with ties to al Qaeda, produces a form of cyanide cream that kills on contact.	William Safire, "Tying Saddam to Terrorist Organizations," *New York Times*, August 25, 2002.
January–June 2002	Biological	Ansar al-Islam has conducted experiments with ricin, a deadly toxin, on at least one human subject.	"US Knew of Bio-Terror Tests in Iraq," BBC News, August 20, 2002; "US Monitors Kurdish Extremists," Fox News, August 21, 2002; Isma'il Zayir, "Ansar al-Islam Group

			Accuses [Jalal] Talabani of Spreading Rumors about Its Cooperation with Al-Qaeda," *Al-Hayah*, August 22, 2002.
January 2002	Nuclear	Diagrams of US nuclear power plants are found in abandoned al Qaeda camps and facilities in Afghanistan.	Bill Gertz, "Nuclear Plants Targeted," *Washington Times*, January 31, 2002; John J. Lumpkin, "Diagrams Show Interest in Nuke Plants," Associated Press, January 30, 2002.
Before March 2002	Biological	US forces discover a BW laboratory under construction near Kandahar that was abandoned by al Qaeda. It was being built to produce anthrax.	Dominic Evans, "US Troops Found Afghan Biological Lab," Reuters, March 22, 2002; Michael R. Gordon, "US Says It Found Qaeda Lab Being Built to Produce Anthrax," *New York Times*, March 23, 2002.
April 2002	Radiological	Abu Zubayda claims that al Qaeda possesses the ability to produce a	Jamie McIntyre, "Zubaydah: al Qaeda Had 'Dirty Bomb' Know-How," CNN,

Time of Incident	WMD	Incident (continued)	Sources (continued)
		radiological weapon and already has one in the United States.	April 22, 2002; "Al-Qaeda Claims 'Dirty Bomb' Know-How," BBC, April 23, 2002.
May 2002	Radiological	US citizen Abdullah al-Muhaji (formerly José Padilla) is arrested in Chicago. He is alleged to have been involved with al Qaeda in a plan to perpetrate a radiological bomb attack in the United States	Dan Eggen and Susan Schmidt, "'Dirty Bomb' Plot Uncovered, US Says: Suspected Al Qaeda Operative Held as 'Enemy Combatant,'" *Washington Post*, June 11, 2002.
May 2002	Chemical	Among the items seized during the arrest of Sami Uthman, a Lebanese national who moved to the United States and became an imam at a Islamist mosque in Seattle, are	Patrick J. McDonnell and Josh Meyer, "Links to Terrorism Probed in Northwest," *Los Angeles Times*, July 13, 2002.

papers by
London-based
al Qaeda
recruiter Shaykh
Abu Hamza
al-Masri,
firearms,
military manuals,
and "instructions
on poisoning
water sources."

June 3, 2002	Radiological	Bin Laden attempts to acquire 11 lbs of radioactive thallium from measuring devices on decommissioned Russian submarines, but Russia's Federal Security Service claims to have blocked the attempt.	"Insider Notes," United Press International, June 3, 2002.
July 18, 2002	Biological	Stephen Younger, director of the Defense Threat Reduction Agency, says that al Qaeda's interest in BWs is focused mainly on anthrax.	"Weapons Worries," CBS News, July 18, 2002.

Time of Incident	WMD	Incident (continued)	Sources
September 13, 2002	Chemical/ Biological	Lab equipment found near Kandahar, Afghanistan, supports the assessment that al Qaeda has acquired the necessary ingredients for "a very limited production of biological and chemical agents."	Judith Miller, "Lab Suggests Qaeda Planned to Build Arms, Officials Say," *New York Times*, September 14, 2002.
October or November 2002	Chemical	The Islamist group Asbat al-Ansar, a Lebanon-based Sunni group affiliated with al Qaeda, reportedly obtained the nerve agent VX from the Iraqi regime.	Barton Gellman, "US Suspects Al Qaeda Got Nerve Agent from Iraqis," *Washington Post*, December 12, 2002.
November 9, 2002	Chemical	British security officials arrest three men, allegedly agents of al Qaeda, planning a cyanide attack	Hala Jaber and Nicholas Rufford, "MI5 Foils Poison-Gas Attack on the Tube," *Sunday Times* (London), November 17,

		on the London subway	2002.
November 2004	Nuclear	Sharif al-Masri, a key al Qaeda operative, informs authorities that bin Laden has arranged to smuggle nuclear supplies and tactical nuclear weapons into Mexico.	"Al Qaeda Wants to Smuggle N-Material to US," *Nation*, November 17, 2004.

Kimberly McCloud, Gary A. Ackerman, and Jeffrey M. Bale, "Chart: Al Qaida's WMD Activities," Center for Nonproliferation Studies, Monterey Institute of International Studies, January 21, 2003. Modified and revised by Paul L. Williams.

NOTES

CHAPTER ONE: THE YOUNG LION AND THE DREAM OF THE AMERICAN HIROSHIMA

1. Roland Jacquard, *In the Name of Osama Bin Laden: Global Terrorism and the Bin Laden Brotherhood* (Durham, NC: Duke University Press, 2002), p. 40.

2. "Usama Bin Laden," FBI's Ten Most Wanted Fugitives, http://www.fbi.gov/mostwant/topten/fugitives/laden.htm.

3. Peter Bergen, *Holy War, Inc.: Inside the Secret World of Osama Bin Laden* (New York: Simon & Schuster, 2002), p. 105.

4. Testimony of L' Hossaine Kherchtou, *The United States v. Osama Bin Laden et alia*, United States District Court, Southern District of New York, February 21, 2001.

5. Testimony of Jamal Ahmed al-Fadl, *The United States v. Osama Bin Laden et alia*, February 14, 2001.

6. Ibid.

7. Gordon Thomas, "Los Malerines de Osama," *El Mundo*, September 23, 2001.

8. Michael Barletta and Erik Jorgensen, "Weapons of Mass Destruction in the Middle East," Center for Nuclear Nonproliferation, Monterey Center for International Studies, Monterey, California, April 1999.

9. Indictment of Mamdouh Mahmud Salim, US Attorney's Office, Southern District of New York, January 6, 1999.

10. Jane Corbin, *Al Qaeda: In Search of the Terror Network That Threatens the World* (New York: Thunder's Mouth Press/Nation Books, 2002), p. 59.

11. "Bin Laden Endorses 'The Nuclear Bomb of Islam,'" *Fact Sheet: The Charges against Osama Bin Laden*, US Department of State, December 15, 1999, http://usinfo.state.gov/topical/pol/terror/99129502.htm. Also see "Interview with Bin Laden," *Time*, December 23, 1998.

12. Rohan Gunaratna, *Inside Al Qaeda: Global Network of Terror* (New York: Berkley Books, 2002), p. 51.

13. Daniel Benjamin and Steven Simon, *The Age of Sacred Terror* (New York: Random House, 2002), pp. 132–33. Also see Simon Reeve, *The New Jackals: Ramzi Yousef, Osama Bin Laden, and the Future of Terrorism* (Boston: Northeastern University Press, 1999), p. 159.

14. Prince Turki bin Faisal, quoted in Corbin, *Al Qaeda*, p. 61.

15. Ibid., pp. 61–62.

16. Bergen, *Holy War, Inc.*, p. 105.

17. Bin Laden, quoted in Corbin, *Al Qaeda*, p. 4.

18. Ibid., p. 7.

19. Jacquard, *In the Name of Osama Bin Laden*, p. 14.

20. Corbin, *Al Qaeda*, p. 7.

21. Reeve, *The New Jackals*, p. 159.

22. The year of Muhammad bin Laden's death remains a matter of dispute. Several biographers, including Daniel Benjamin and Steven Simon, say that he died in 1967. Others, such as Yossef Bodansky, argue that his death occurred in 1973.

23. Jacquard, *In the Name of Osama Bin Laden*, p. 12.

24. Benjamin and Simon, *The Age of Sacred Terror*, p. 96; Bergen, *Holy War, Inc.*, p. 50.

25. Ibid., p. 96.

26. Wayne Madsen, "Questionable Ties," *In These Times*, 2002. http://www.inthesetimes.com/issue/25/25/feature3.shtml.

27. Berkeley Rice, *Trafficking: The Boom and Bust of the Air America Cocaine Ring* (New York: Scribners, 1989), pp. 24–32.

28. Ibid.

29. Andrew Wheat, "The Bush Bin–Laden Connection," *Texas Observer*, November 9, 2001.

30. Ibid.

31. Gunaratna, *Inside Al Qaeda*, p. 26.

32. Reeve, *The New Jackals*, p. 160.

33. Corbin, *Al Qaeda*, p. 8.

34. Paul L. Williams, *Al Qaeda: Brotherhood of Terror* (Indianapolis: Alpha Books, 2002), p. 75.

35. Reeve, *The New Jackals*, p. 161.

36. Bin Laden, quoted in Bergen, *Holy War, Inc.*, p. 55.

37. Ibid.

38. Milt Beardon, quoted in Corbin, *Al Qaeda*, p. 16.

39. Jacquard, *In the Name of Osama Bin Laden*, pp. 22–23.

40. Robert Young Pelton, *The World's Most Dangerous Places*, 5th ed. (New York: HarperCollins, 2002), p. 347.

41. Williams, *Al Qaeda: Brotherhood of Terror*, p. 76.

42. Ibid., pp. 327–28.

43. Ibid., p. 328.

44. Lawrence Wright, "The Man behind Bin Laden: How an Egyptian Doctor Became a Master of Terror," *New Yorker*, September 16, 2002.

45. Williams, *Al Qaeda: Brotherhood of Terror*, p. 78.

46. Ibid.

47. Ibid.

48. Corbin, *Al Qaeda*, pp. 18–19.

49. Ibid.

50. Benjamin and Simon, *The Age of Sacred Terror*, p. 103.

51. Gunaratna, *Inside Al Qaeda*, p. 31.

52. Sayyid Qutb, quoted in Nazih N. Ayubi, *Political Islam: Religion and Politics in the Arab World* (London: Routledge, 1991), p. 140.

53. Bin Laden, quoted in Corbin, *Al Qaeda*, p. 21.

54. Testimony of Jamal Ahmed al-Fadl, *The United States of America v. Osama bin Laden, et alia*, February 6, 2001.

55. Ibid.

56. "Who's Who in Al Qaeda," BBC News, February 19, 2003. Also see "Status of Major Al Qaeda Members," Fox News, March 1, 2003.

57. Bin Laden, quoted in Reeve, *The New Jackals*, p. 170.

58. Jane Corbin, *Al Qaeda*, p. 27.

59. Ibid.

60. Osama bin Laden, "Jihad against Jews and Christians," World Islamic Statement, February 23, 1998, appendix B in *Osama's Revenge: The Next 9/11—What the Media and the Government Haven't Told You*, by Paul L. Williams (Amherst, NY: Prometheus Books, 2004), p. 215.

61. Benjamin and Simon, *The Age of Sacred Terror*, pp. 107–108.

62. Reeve, *The New Jackals*, p. 172.

63. Ibid.

64. Ibid., p. 173.

65. Bergen, *Holy War, Inc.*, pp. 86–87.

66. Benjamin and Simon, *The Age of Sacred Terror*, p. 135.

67. Abdel Bari Atwan, quoted in ibid., p. 96.

CHAPTER TWO: THE GOOD LIFE AMONG THE TALIBAN

1. Jane Corbin, *Al Qaeda: In Search of the Terror Network That Threatens the World* (New York: Thunder's Mouth Press/Nation Books, 2002), pp. 65–66.

2. Simon Reeve, *The New Jackals: Ramzi Yousez, Osama Bin Laden, and the Future of Terrorism* (Boston: Northeastern University Press, 1999), p. 190.

3. Roland Jacquard, *In the Name of Osama Bin Laden: Global Terrorism and the Bin Laden Brotherhood* (Durham, NC: Duke University Press, 2002), p. 41.

4. Robert Young Pelton, *The World's Most Dangerous Places*, 4th ed. (New York: HarperResource, 2000), p. 284.

5. Daniel Benjamin and Steven Simon, *The Age of Sacred Terror* (New York: Random House, 2002), p. 135.

6. Corbin, *Al Qaeda*, p. 64.

7. "Deobandi Islam: The Religion of the Taliban," Defense Language Institute, Global Security Network, April 2001, http://globalsecurity.org/military/library/report/2001/Deobandi_Islam.pdf.

8. Ibid.

9. Corbin, *Al Qaeda*, p. 65.

10. Ibid.

11. Reeve, *The New Jackals*, p. 190.

12. Ibid., p. 191.

13. Paul L. Williams, *Osama's Revenge: The Next 9/11—What the Media and the Government Haven't Told You* (Amherst, NY: Prometheus Books, 2004), p. 26.

14. Najibullah's communiqué in "Deobandi Islam."

15. Benjamin and Simon, *The Age of Sacred Terror*, pp. 136–37.

16. Ibid.

17. Williams, *Osama's Revenge*, p. 27.

18. Jacquard, *In the Name of Osama Bin Laden*, p. 42.

19. Pelton, *The World's Most Dangerous Places*, p. 42.

20. Abu Abdul Aziz al-Afghani, quoted in Yossef Bodansky, *Bin Laden: The Man Who Declared War on America* (New York: Forum, 2001), p. 189.

21. Peter Bergen, *Holy War, Inc.: Inside the Secret World of Osama Bin Laden* (New York: Simon & Schuster, 2002), p. 97.

22. Osama Bin Laden, "Declaration of War against the Americans Occupying the Land of the Two Holy Places," complete text printed as appendix A in Williams, *Osama's Revenge*, pp. 179–213.

23. Yonah Alexander and Michael S. Swetnam, *Usama Bin Laden's Al Qaida: Profile of a Terror Network* (Ardsley, NY: Transnational Publishers, 2001), p. 30.

24. Bergen, *Holy War, Inc.*, p. 106.

25. Joel Mowbray, "Saudis Behaving Badly," *National Review*, December 20, 2002.

26. Ibid.

27. Dick Gannon, quoted in Reeve, *The New Jackals*, p. 207.

28. Benjamin and Simon, *The Age of Sacred Terror*, p. 144.

29. Mowbray, "Saudis Behaving Badly."

30. Benjamin and Simon, *The Age of Sacred Terror*, p. 139.

31. Jacquard, *In the Name of Osama Bin Laden*, p. 48.

32. Ibid.

CHAPTER THREE: FROM ALBANIA TO THE ATOM BOMB

1. Michael Paterchick, quoted in "New Jersey Heroin Is Purest in U.S.," *Newark Star Ledger*, December 12, 2004.

2. Colin Freeman, "Afghanistan's Disturbing Poppy Explosion: U.N. Says Nation Tops Colombia as Capital of Illicit Narcotics," *San Francisco Chronicle*, November 19, 2004.

3. Ibid.

4. Terry Frieden, "FBI: Albanian Mobsters 'New Mafia,'" CNN, August 19, 2004.

5. Ibid.

6. Jerry Capeci, "Zef's Got Staying Power Too," *Gangland*, September 4, 2003, http://www.ganglandnews.com/column346.html#zef.

7. Mike Brunker, "Alleged Mobsters Guilty in Vast Net, Phone Fraud," MSNBC, February 15, 2005.

8. Gus Xhudo, "Men of Purpose: The Growth of Albanian Criminal Activity," Ridgeway Center for International Security Studies, University of Pittsburgh, Spring 1996.

9. M. Bozinovich, "The New Islamic Mafia," *Serbianna*, February 21, 2005, http://www.serbianna.com/columns/mb/o28.shtml.

10. Anthony M. DeStefano, "The Balkan Connection," *Wall Street Journal*, September 9, 1985.

11. Ibid.

12. Bozinovich, "The New Islamic Mafia."

13. Cataldo Motta, quoted in ibid.

14. Dusan Janjic, quoted in ibid.

15. Agim Gashi, quoted in Roberto Ruscica, "The Albanian Mafia: This Is How It Helps the Kosovo Government," *Corriere della Sera* (Milan), October 17, 1998. Also see William Norman Grigg, "Diving into the Kosovo Quagmire," *New American*, March 15, 1999, http://www.thenewamerican .com/tna/1999/03-15-99/kosovo .html.

16. Gashi, quoted in Ruscica, "The Albanian Mafia: This Is How It Helps the Kosovo Government."

17. Marcia Christoff Kurop, "Al Qaeda's Balkan Links," *Wall Street Journal*, November 1, 2001.

18. Daniel Benjamin and Steven Simon, *The Age of Sacred Terror* (New York: Random House, 2002), pp. 146–47. Also see "Bin Laden: A Nuclear Threat," EERI Daily Intelligence Report, November 7, 2001, http://www .emergencycom/2001.

19. Kurop, "Al Qaeda's Balkan Links."

20. Claire Sterling, *Octopus: How the Long Reach of the Sicilian Mafia Controls the Global Narcotics Trade* (New York: Simon & Schuster, 1990), pp. 82–84.

21. Ibid., p. 162.

22. Umberto Pascali, "KLA and Drugs: The 'New Colombia of Europe' Grows in the Balkans," *Executive Intelligence Review*, June 22, 2001.

23. Stella L. Jatras, "The Crimes of the KLA: Who Will Pay?" Anti-War News, March 14, 2002, http://www.antiwar.com/article/php?articleid=1499.

24. Ibid.

25. Ibid.

26. Paul L. Williams, *Al Qaeda: Brotherhood of Terror* (Indianapolis: Alpha Books, 2002), p. 165.

27. Ibid.

28. Research Analysis, Centre of Peace in the Balkans, May 2000, http://www.balkanpeace.org/our/our02.shtml.

29. Ibid.

30. Tony White, quoted in ibid.

31. Paul L. Williams, *Osama's Revenge: The Next 9/11—What the Media and the Government Haven't Told You* (Amherst, NY: Prometheus Books, 2004), p. 33.

32. Jason Burke, "Afghanistan: Heroin in the Holy War," *New Delhi Observor*, December 6, 1998.

33. Fareed Zakaria, "The New Rules of Engagement," *Newsweek*, December 6, 2001.

34. "National Household Survey of Drug Abuse," US Department of Health and Human Services, 2000, http://www.oas.samhsa.hov/nhsda.htm#NKSDAinfo.

35. Jason Bartholet and Steve Levine, "The Holy Men of Heroin," *Newsweek*, December 6, 1999.

36. Ibid.

37. Roland Jacquard, *In the Name of Osama Bin Laden: Global Terrorism and the Bin Laden Brotherhood* (Durham, NC: Duke University Press, 2002), pp. 42–43.

38. Bartholet and Levine, "The Holy Men of Heroin."

39. Jacquard, *In the Name of Osama Bin Laden*, p. 138.

CHAPTER FOUR: THE THREE WARS

1. Peter Wolf, "The Assassination of Ahmad Shah Massoud," Center for Research on Globalization, September 14, 2003, http://www.global research.co/articles/wol309A.html.

2. Ibid. Also see "Albanian Militants' Training Camp Operates in U.S.-Controlled Kosovo," *Pravda*, October 16, 2001.

3. Tom Walker and Aiden Laverty, "CIA Aided Kosovo Guerrilla Army," *Sunday Times* (London), March 12, 2000.

4. Jon Silberman, "Racak Massacre Haunts Milosevic Trial," BBC News, February 12, 2002.

5. "Kosovo: Obscure Areas of a Massacre," *Le Figaro* (Paris), January 20, 1999. Also see "Were the Racak Dead Really Coldly Massacred?" *Le Monde* (Paris), January 20, 1999.

6. "Kosovo Fact-Finding Mission," A White Paper of the Religious Freedom Coalition, August 2004, http://www.serbianunity.net/bydate/2004/October 9/file/1096632196_3t3tqqcixlx.kosovowhitepaper.html.

7. Ibid.

8. Robert Fisk, "Serbs Murdered by the Hundreds," *Independent*, November 24, 1999.

9. David Lynch, "Serbs Fear They Will Be Eliminated from Kosovo," *USA Today*, March 27, 2000.

240   NOTES

10. Michael Radu, "NATO's KLA Problem," Foreign Policy Research Institute, September 8, 1999, http://www.fri.org/enotes/balkansturkey .19990908.radu.klaproblem.html.

11. Ibid. Also see Lynch, "Serbs Fear They Will Be Eliminated from Kosovo."

12. "Jihad against Jews and Crusaders," World Islamic Front statement, February 23, 1998, in *Usama Bin Laden's Al Qaeda: Profile of a Terrorist Network*, ed. Yonah Alexander and Michael S. Swetnam (Ardsley, NY: Transnational Publishers, 2001).

13. Michael Grunwald, "CIA Helps Thwart Bomb Plot against Embassy in Uganda," *Seattle Times*, September 25, 1998.

14. Rohan Gunaratna, *Inside Al Qaeda: Global Network of Terror* (New York: Berkley Books, 2002), p. 63.

15. Peter L. Bergen, *Holy War, Inc.: Inside the Secret World of Osama Bin Laden* (New York: Simon & Schuster, 2002), p. 125.

16. Ibid.

17. Mauvi Fazlur Rehman Khalil, quoted in Yossef Bodansky, *Bin Laden: The Man Who Declared War on America* (New York: Forum, 1999), pp. 284–85.

18. Rahimullah Yusufzai, quoted in ibid., p. 295.

19. Hassan Abdullah al-Turabi, quoted in ibid.

20. Jane Corbin, *Al Qaeda: In Search of the Network That Threatens the World* (New York: Thunder's Mouth Press/Nation Books, 2003), p. 95.

21. Paul L. Williams, *Osama's Revenge: The Next 9/11—What the Media and the Government Haven't Told You* (Amherst, NY: Prometheus Books, 2004), pp. 61–62.

22. Ibid., pp. 63–64.

23. Daniel Pipes, *Militant Islam Reaches America* (New York: Norton, 2002), p. 247.

24. "Presidential Address to the Nation," September 20, 2001, White House Press Service, Washington, DC.

25. Luke Harding and Jason Burke, "U.S. Blamed for 100 Missile Deaths," *Guardian* (Islamabad), October 12, 2001.

26. Corbin, *Al Qaeda*, p. 260.

27. "Al Qaeda's Nuclear Plans Confirmed," BBC News, November 16, 2001, http://news.bbc.co.uk/1/hi/world/south_asia/1657901.stm. Also see Kathy Gannon, "Plans Reveal Al Qaeda's Nuclear Goals," *Tulsa World*, November 16, 2001.

28. David Albright, quoted in Graham Allison, *Nuclear Terrorism: The*

Ultimate Preventable Catastrophe (New York: Times Books/Henry Holt and Company, 2004), p. 26.

29. During the first week of the Afghan offensive, M.Sgt. Evander Earl Andrews was killed in a heavy-equipment accident. The accident occurred in the Arabian Peninsula. Although Andrews was not killed in combat, his death is often listed as the first casualty of Operation Enduring Freedom.

30. Antoine Blua, "Al Qaeda Fighters Routed from Tora Bora," *Radio Free Europe*, December 11, 2001, http://www.rfel.org/features/2001/12/11122001082737.asp.

31. David Ensor, "Close In on Al Qaeda," CNN, December 21, 2001.

32. Barbie Dutter and Ben Fenton, "Uranium and Cyanide Found in Drums at Bin Laden's Base," *Daily Telegraph* (London), December 24, 2001.

33. Williams, *Osama's Revenge*, p. 75. Also see "Bin Laden Sought on Christmas," United Press International, December 24, 2001, http://www.newsmax.com/archives/articles/2001/12/24/202349.shtml.

34. "Al Qaeda's Nuclear Plans Confirmed."

35. "Operation Anaconda: Questionable Outcomes for the United States," Global Intelligence Company, March 11, 2002, http://www.stratfor.com/fib/fib_view.php?ID=203443.

36. President Bush, quoted in Mushahid Hussein, "Operation Anaconda: Win-Win, Lose, Lose," *Asia Times* March 22, 2002.

37. Barry Bearak, "Details of Victory Are Unclear, but It Is Celebrated, Nonetheless," *New York Times*, March 14, 2002.

38. Vivienne Walt, "No Bodies Where Battle Began," *USA Today*, March 14, 2002.

39. Bearak, "Details of Victory Are Unclear."

40. James Gordon Meek, "Officials Fear Al Qaeda Nuclear Attack," *New York Daily News*, March 14, 2003.

Chapter Five: The Loose Nukes

1. Robert Young Pelton, *The World's Most Dangerous Places*, 4th ed. (New York: HarperResource, 2000), p. 783.

2. Ibid., p. 782.

3. Ibid.

4. Ibid., p. 783.

5. Graham Allison, *Nuclear Terrorism: The Ultimate Preventable Catastrophe* (New York: Times Books/Henry Holt and Company, 2004), p. 70.

6. Pelton, *The World's Most Dangerous Places*, p. 783.

7. Dick Cheney, quoted in Allison, *Nuclear Terrorism*, p. 61.

8. William C. Potter, "Nuclear Profiles of the Soviet Successor States," Center for Nonproliferation Studies, Monterey Institute of International Studies, Monterey, California, September 14, 1993.

9. Rensselaer Lee, *Smuggling Armageddon: The Nuclear Black Market in the Former Soviet Union and Europe* (New York: St. Martin's, 1998), pp. 135–36.

10. Oleg Bukharin and William Potter, "Potatoes Were Guarded Better," *Bulletin of the Atomic Scientists* (May–June 1995).

11. Phil Williams and Paul Woessner, "The Real Threat of Nuclear Smuggling," *Scientific American* (January 1996).

12. "Chechnya's Special Weapons," Global Security, October 2001, http://www.globalsecurity.org/wmd/wond/chechnya/.

13. Ibid.

14. Williams and Woessner, "The Real Threat of Nuclear Smuggling."

15. Lee, *Smuggling Armageddon*, p. 136.

16. Andrew Cockburn and Scott Cockburn, *One Point Safe* (New York: Doubleday, 1997), pp. 101–103.

17. David Smigielski, "A Review of the Nuclear Suitcase Bomb Controversy," Policy Update, Russian-American Nuclear Security Council (RANSAC), September 2003, http://www.ransac.org/Documents/suitcase nukes/090103.pdf.

18. Alexander Lebed, quoted in Scott Parish, "Are Suitcase Nukes on the Loose? The Story behind the Controversy," Center for Nonproliferation Studies, Monterey Institute of International Studies, November 1997.

19. Ibid.

20. Ibid.

21. "The Perfect Terrorist Weapon," interview with Gen. Alexander Lebed, *60 Minutes*, CBS News, September 7, 1997. Portion of interview also contained in Allison, *Nuclear Terrorism*, pp. 43–45.

22. "Suitcase Nukes: A Reassessment," Center for Nonproliferation Studies, Monterey Institute of International Studies, September 2002.

23. Allison, *Nuclear Terrorism*, p. 49.

24. Ibid.

25. Ibid. Also see Stephen Schwartz, ed., *Atomic Audit: The Costs and Consequences of U.S. Nuclear Weapons Since 1940* (New York: Brookings Institution Press, 1998), p. 49.

26. Viktor Chernomyrdin, quoted in Scott Parish, "Are Suitcase Nukes on the Loose?"

27. *Rossiyskaya Gazeta*, quoted in ibid.

28. Ibid.

29. Ibid. Also see Paul Williams, *Osama's Revenge: The Next 9/11—What the Media and the Government Haven't Told You* (Amherst, NY: Prometheus Books, 2004), pp. 46–47.

30. Parish, "Are Suitcase Nukes on the Loose?"

31. Ibid.

32. Allison, *Nuclear Terrorism*, pp. 48–49.

33. Carey Sublette, "Are Suitcase Bombs Possible?" Nuclear Weapons Archive, May 18, 2001, http://nuclearweaponarchive.org/News/DoSuitcase NukesExist.html.

34. Nikolai Sokov, "Tactical Nuclear Weapons," Center for Nonprolif-eration Studies, Monterey Institute of International Studies, May 2002.

35. Rep. Curt Weldon, quoted in Pamela Hess, "Experts Doubt Al Qaeda Nuclear Claims," United Press International, March 3, 2004.

36. Marie Calvin, "Holy War with U.S. in His Sights," *Times* (London), August 16, 1998.

37. Rohan Gunaratna, *Inside Al Qaeda: Global Network of Terror* (New York: Berkley Books, 2004), p. 181.

38. Ibid.

39. Author's interview with Ms. Whitcraft, September 14, 2004.

40. "Osama's Nukes Traced to Soviet General," *Newsmax.com*, July 18, 2004, http://www.newsmax.com/archieves/articles/2004/7/18/213209.shtml.

41. Greta Van Sustern's interview with Hamid Mir, *On the Record*, Fox News, March 22, 2002.

42. Yossef Bodansky, quoted in "Bin Laden Has 20 Nuclear Bombs," *World Tribune* (Washington, DC), August 9, 1999.

43. Shaykh Hisham Kabbani, quoted in "Bin Laden Endorses 'The Nuclear Bomb of Islam,'" *Fact Sheet: The Charges against Osama Bin Laden*, US Department of State, December 15, 1999, http://usinfo.state.gov/topical/pol/terror/99129502.htm.

44. Williams, *Osama's Revenge*, p. 165.

45. Stewart Stogel, "Author: Al-Qaida Has Nuclear Weapons Inside U.S.," *Newsmax*, July 14, 2004, http://www.newsmax.com/archives/articles/2004/7/13/171536.shtml.

46. "N-Weapons May Be in US Already," *Daily Telegraph* (Sydney, Aus-tralia), November 14, 2001. Also see Naveed Miraj, "Al Qaeda Nukes May Already Be in the US," *Frontier Post* (Islamabad), November 11, 2001. The Miraj piece contains verification from Pakistan's ISI.

47. "N-Weapons May Be in US Already."

48. Richard Sale, "Feds Look for Smuggled Nukes in the United States," United Press International, *Newsmax*, December 21, 2001, http://www.newsmax.com/archives/articles/2001/12/20/181037.shtml.

49. Ibid.

50. George Tenet, quoted in Williams, *Osama's Revenge*, p. 168.

51. "Bin Laden Endorses 'The Nuclear Bomb of Islam.'"

52. "Interview with Bin Laden," *Time*, December 23, 1998.

53. John Miller's interview with Osama Bin Laden, ABC News, May 24, 1998, http://abcnews.go.com/sections/world/DailyNews/Miller_binladen_980609.html.

54. Hamid Mir, "If US Uses Nuclear Weapons, It Will Receive Same Response: Interview with Osama Bin Laden," *Dawn* (Pakistan), November 10, 2001. See also Evan Thomas, "The Possible Endgame and Future of Al Qaeda," *Newsweek*, November 26, 2001.

55. "Bin Laden's Nuclear Weapons," *Insight*, November 2, 2001.

56. Richard Sale, "Israel Finds Radiological Backpack Bomb," United Press International, October 14, 2001, http://www.papillonartpalace.com/israel.htm.

57. Richard Sale, "Feds Look for Smuggled Nukes in the United States," United Press International, *Newsmax*, December 21, 2001, http://www.newsmax.com/archives/articles/2001/12/20/181037.shtml.

58. *Insight*, December 21, 2001.

CHAPTER SIX: THE FIVE-YEAR INTERMISSION

1. Igor Valynkin, quoted in Scott Parrish, "Suitcase Nukes: A Reassessment," Center for Nonproliferation Studies, Monterey Institute of International Studies, Monterey, California, September 23, 2003.

2. Ibid.

3. Roland Jacquard, *In the Name of Osama Bin Laden* (Durham, NC: Duke University Press, 2002), p. 142.

4. Testimony of Col. Stanislav Lunev, National Security Committee, *Hearing on Russian Threats*, January 2000.

5. Ibid.

6. Yossef Bodansky, *Bin Laden: The Man Who Declared War on America* (New York: Forum, 1999), p. 330.

7. Peter L. Bergen, *Holy War, Inc.: Inside the Secret World of Osama Bin Laden* (New York: Simon & Schuster, 2002), p. 242.

8. Paul Williams, *Al Qaeda: Brotherhood of Terror* (Indianapolis: Alpha Books, 2002), p. 10.

9. Daniel Benjamin and Steven Simon, *The Age of Sacred Terror* (New York: Random House, 2002), p. 393.

10. Vincent Morris, "Pol Fears Soviets Hid A-Bombs Across U.S.," *New York Post*, November 7, 1999.

11. Congressman Curt Weldon, quoted in ibid.

12. "FBI Director Admits Russians May Have Secret Weapons," *Newsmax*, November 8, 1999, http://www.newsmax.com/articles/?a=1999/11/8/73601.

13. Testimony of Vasili Mitrakhin, House Committee on Government Reform, *Russian Threats to US Security in the Post–Cold War Era*, 106th Cong., 6th sess., January 4, 2000.

14. Testimony of Col. Stanislav Lunev, House Committee, *Russian Threats.* Col. Lunev elaborates on the buried nukes in his book *Through the Eyes of the Enemy* (New York: Brassey, 2002).

15. Morris, "Pol Fears Soviets Hid A-Bombs Across U.S."

16. Michael Crowley, "Can Terrorists Build the Bomb?" *Popular Science*, March 3, 2004.

17. Ibid.

18. Senator Joseph F. Biden, quoted in Graham Allison, *Nuclear Terrorism: The Ultimate Preventable Catastrophe* (New York: Times Books, 2004), p. 95.

19. Ibid.

20. Ibid., p. 93.

21. "Bin Laden's Search for Nuclear Weapons," *US News & World Report*, October 5, 1998.

22. Jacquard, *In the Name of Osama Bin Laden*, p. 144.

23. Ryan Mauro, "Terrorist Possession of Weapons of Mass Destruction," *World Threats*, Monthly Analysis, February 2003, http://www.worldthreats.vcom/monthly%20Analysis/MA%202003.htm. Also see Robert Friedman, "The Most Dangerous Mobster in the World," *Village Voice*, May 22, 1998.

24. Mauro, "Terrorist Possession of Weapons of Mass Destruction."

25. Ryan Mauro, "The Next Attack on America," *World Threats*, November 27, 2003, http://www.freepublic.com/focus/f-news/1020690/posts. Also see "Bin Laden Buys Nuclear Materials," *World Net Daily*, November 26, 2003.

26. Paul L. Williams, *Osama's Revenge: The Next 9/11—What the Media*

and the Government Haven't Told You (Amherst, NY: Prometheus Books, 2004), p. 75.

CHAPTER SEVEN: ENTER DR. EVIL

1. John J. Curtis, "Pakistan's Bomb Maker," *Online Columnist.com*, January 5, 2003, http://www.onlinecolumnist.com/01503.htm.

2. "Profile: Abdul Qadeer Khan," BBC News, December 20, 2003, http://news.bbc.co.uk/1/hi/world/south_asia/3343621.

3. Jeffrey Goldberg, "Inside Jihad U.: The Education of a Holy Warrior," *New York Times*, June 25, 1998.

4. Ibid.

5. Bernard-Henri Levy, "Pakistan Must Provide Proof of Reforming the ISI," *South Asia Tribune*, September 14, 2003.

6. Rajesh Kumar Mishra, "Pakistan as a Proliferator State: Blame It on Dr. A. Q. Khan," South Asia Analysis Group, paper no. 567, December 20, 2002, http://www.saag.org/papers6/paper567.html.

7. Rajesh Kumar Mishra, "Nuclear Scientific Community of Pakistan: Clear and Present Danger to Nonproliferation," South Asia Analysis Group, paper no. 601, July 2, 2003, http://www.saag.or/papers7/paper601.html.

8. Zaffar Abbas, "US Bans Trade with Pakistani Nuclear Lab," *Guardian* (Islamabad), April 2, 2003.

9. Curtis, "Pakistan's Bomb Maker."

10. Joe Trento, "Pakistan and Iran's Scary Alliance," *Public Education Center*, National Security and Natural Resources News Services, August 15, 2003, http://www.publicedcenter.org/stories/trento/2003-8-15.

11. John J. Curtis, "Why Iraq?" *OnlineColumnist.com*, January 28, 2003, http://www.onlinecolumnist.com/012803.html.

12. Arnaud de Borchgrave, "Pakistan's Paranoid Panjandrum," *Washington Times*, January 20, 2003.

13. Ibid. Also see Maggie Farley and Bob Drogan, "The Evil behind the Axis," *Los Angeles Times*, January 5, 2003.

14. Matt Kelley, "Libyan Success Exposes Bush Problem: Pakistan," *Scranton (PA) Tribune*, January 14, 2004.

15. Johanna McGreary, "Inside the A-Bomb Bazaar," *Time*, January 19, 2004.

16. Michael R. Gordan, "Giving Up Those Weapons: After Libya, Who Is Next?" *New York Times*, January 1, 2004.

17. William J. Broad and David E. Sanger, "The Bomb Merchant: Chasing Dr. Khan's Network; As Nuclear Secrets Emerge, More Are Suspected," *New York Times*, December 26, 2004.

18. Ibid.

19. Barton Gellman and Dafna Linzer, "Unprecedented Peril Forces Tough Calls," *Washington Post*, October 27, 2004.

20. Ibid.

21. Mohan Malik, "A. Q. Khan's China Connection," Jamestown Foundation, May 8, 2004.

22. Mishra, "Pakistan as a Proliferator State."

23. B. Rahman, "Lashkar-e-Toiba: Its Past, Present, and Future," Institute for Topical Studies, Chennai, India, December 25, 2000.

24. Ibid.

25. Ibid.

26. Ben Fenton and Ahmid Rashid, "US Hails Capture of Bin Laden's Deputy," *Daily Telegraph* (Sydney), February 4, 2002.

27. Eric Stakelbeck, "Terror's South American Front," *Front Page*, March 19, 2004, http://www.frontpagemag.com/Articles/ReadArticle.asp?ID=12643.

28. Kaushik Kapisthalam, "Pakistan's Forgotten Al-Qaeda Nuclear Link," *Asia Times*, June 3, 2004.

29. Mishra, "Pakistan as a Proliferator State." Also see Kaushik Kapisthalam, "Outside View: No Free Passes for Pakistan," *Washington Times*, December 30, 2003.

30. Peter L. Bergen, *Holy War, Inc.: Inside the Secret World of Osama Bin Laden* (New York: Simon & Schuster, 2002), p. 244.

31. Peter Baker, "Pakistani Scientist Who Met Bin Laden Failed Polygraphs, Renewing Suspicion," *Washington Post*, March 3, 2002.

32. Ibid. Also see Daniel Benjamin and Steven Simon, *The Age of Sacred Terror* (New York: Random House, 2002), pp. 203–204.

33. Rifaat Hussein, quoted in Baker, "Pakistani Scientist Who Met Bin Laden Failed Polygraphs, Renewing Suspicion."

34. Ibid.

35. Ibid. Also see Benjamin and Simon, *The Age of Sacred Terror*, pp. 203–204.

36. Baker, "Pakistani Scientist Who Met Bin Laden Failed Polygraphs, Renewing Suspicion."

37. Dr. Sultan Bashiruddin Mahmood, quoted in Robert Sam Anson, "The Journalist and the Terrorist," *Vanity Fair* (August 2002).

38. Benjamin and Simon, *The Age of Sacred Terror*, pp. 203–204.

39. Anson, "The Journalist and the Terrorist."

40. Ibid.

41. Baker, "Pakistani Scientist Who Met Bin Laden Failed Polygraphs, Renewing Suspicion."

42. Ibid.

43. Kapisthalam, "Pakistan's Forgotten Al-Qaeda Nuclear Link."

44. Ibid.

45. Borchgrave, "Al Qaeda Nuclear Agenda Verified," *Washington Times*, December 10, 2001.

46. Ibid.

47. Tim Burger and Tim McGirk, "Al Qaeda's Nuclear Contact," *Time*, May 12, 2003.

48. James Gordon Meek, "Officials Fear Al Qaeda Nuke Attack," *New York Daily News*, March 14, 2003.

49. Levy, "Pakistan Must Provide Proof of Reforming the ISI."

50. Nigel Hawkes, "The Nuclear Threat: Pakistan Could Lose Control of Its Nuclear Arsenal," *Times* (London), September 20, 2001.

51. Musharraf, quoted in Mishra, "Pakistan as a Proliferator State."

52. Ash-har Quraishi, "U.S. Supports Nuclear Pardon," CNN, February 5, 2004.

53. Ibid.

54. McGreary, "Inside the A-Bomb Bazaar."

55. Ibid.

56. Douglas Jehl, "CIA Says Pakistanis Gave Iran Nuclear Aid," *New York Times*, November 24, 2004.

57. Broad and Sanger, "The Bomb Merchant."

58. Ibid.

59. Ibid.

60. Jack Pritchard, quoted in ibid.

61. Nicholas D. Kristof, "Twisting Dr. Nuke's Arm," *New York Times*, September 27, 2004.

62. Pakistani president Pervez Musharraf, quoted in B. Rahman, "A. Q. Khan and Osama Bin Laden," *Kashmir Herald* 3, no. 10 (March–April 2004), http://www.kashmirherald.com/featuredarticle/khanandbinladen.html.

63. Ibid.

64. Ibid.

65. Graham Allison, *Nuclear Terrorism: The Ultimate Preventable Catastrophe* (New York: Times Books/Henry Holt and Company, 2004), pp. 61–63.

66. Robert Gallucci, quoted in Seymour Hersh, "The Deal: Why Is Washington Going So Easy on Pakistan's Nuclear Black Marketers," *New Yorker*, March 3, 2004.

CHAPTER EIGHT: WELCOME, OSAMA, TO SOUTH AMERICA

1. Peter Hudson, "There Are No Terrorists Here," *Newsweek*, November 19, 2001.
2. Ibid.
3. "Terrorist and Organized Crime Groups in the Tri-Border Area of South America," Federal Research Division, Library of Congress, under an Interagency Agreement with the director of the Central Intelligence Crime and Narcotics Center, July 2003.
4. Ibid.
5. Ibid.
6. Lt. Col. Philip K. Abbot, "Terrorist Threat in the Tri-Border Area," *Military Review* (September–October 2004).
7. Mark S. Steinitz, "Middle East Terrorist Activity in Latin America," *Policy Papers on the Americas*, vol. 14, study 7 (Washington, DC: Center for Strategic and International Studies, July 2003).
8. Robert Young Pelton, *The World's Most Dangerous Places*, 5th ed. (New York: HarperCollins, 2000), p. 389.
9. William W. Mendel, "Paraguay's Ciudad del Este and the New Centers of Gravity," *Military News* (March–April 2002).
10. Hudson, "There Are No Terrorists Here."
11. Ibid.
12. Pelton, *The World's Most Dangerous Places*, p. 389.
13. Martin Edwin Andersen, "Al Qaeda Across America," *Insight*, November 2, 2001.
14. Ibid.
15. Henry Chu, "Terrorist Suspicions Persist at Border Town in Brazil," *Los Angeles Times*, December 26, 2004.
16. Erick Stakelbeck, "Terror's South America Front," *Front Page*, March 19, 2004, http://www.frontpagemag.com/articles/2004-03-19.
17. Rohan Gunaratna, *Inside Al Qaeda: Global Network of Terror* (New York: Berkley Books, 2002), p. 221.
18. Harris Whitbeck and Ingrid Arneson, "Terrorists Find Haven in

South America," CNN, November 7, 2001; Jeffrey Goldberg, "In the Party of God," *New Yorker*, October 10, 2001.

19. Goldberg, "In the Party of God."

20. Ibid.

21. Ibid.

22. Magnus Ranstrop, quoted in Mike Boettcher, "U.S. Renews Bid to Catch Beirut Bombing Suspect," CNN, October 10, 2001.

23. Yael Shahar, "Al-Qaida's Links to Iranian Security Services," International Policy Institute for Counterterrorism, Herzlia, Israel, January 20, 2003.

24. Ibid.

25. Ibid.

26. Dana Priest and Douglas Farah, "Terror Alliance Has U.S. Worried: Hezbollah, Al Qaeda Seen Joining Forces," *Washington Post*, June 30, 2002.

27. LCpl. John R. Lawson III, "Barracks Bombing, 18 Years Ago," Press Release, US Marine Corps Headquarters, October 18, 2001.

28. Ibid.

29. Yossef Bodansky, *Bin Laden: The Man Who Declared War on America* (New York: Random House, 1999), p. 34.

30. Lenny Ben-David, "Sunni and Shiite Terrorist Networks: Competition or Collusion," *Jerusalem Issue Brief* (Jerusalem Center for Public Affairs) 2, no. 12, (December 18, 2003).

31. Bodansky, *Bin Laden*, p. 38.

32. Lenny Ben-David, "Sunni and Shiite Terrorist Networks."

33. Ibid.

34. Gunaratna, *Inside Al Qaeda*, pp. 196–97.

35. "Main Events in Hamas' History," *USA Today*, March 3, 2004.

36. Matthew Levitt, "New Arena for Iranian-Sponsored Terrorism: The Arab-Israeli Heartland," *Policy Watch*, February 22, 2002.

37. Priest and Farah, "Terror Alliance Has U.S. Worried."

38. Bodansky, *Bin Laden*, p.157.

39. Ibid., p. 158.

40. Aaron Mannes, "Terrorism's Godfather," *National Review*, November 11, 2004.

41. Shahar, "Al-Qaida's Links to Iranian Security Services."

42. Dr. J. R. Albani, "The Syria-Al Qaida Connection," Lebanese Foundation for Peace, May 24, 2003.

43. Mannes, "Terrorism's Godfather."

44. Paul L. Williams, *Osama's Revenge: The Next 9/11—What the Media*

and the Government Haven't Told You (Amherst, NY: Prometheus Books, 2004), p. 126; Faye Bowers, "Iran Holds Top Al Qaeda Leaders," *Christian Science Monitor*, July 28, 2003.

45. Ilene R. Prusher and Philip Smucker, "Al Qaeda Quietly Slipping into Iran, Pakistan," *Christian Science Monitor*, January 14, 2002.

46. Jack Kelly, "What the Terrorists Understand But Pols and Journos Don't," *Jewish World Review*, September 15, 2003.

47. Joe Trento, "Pakistan and Iran's Scary Alliance," *Public Education Center*, National Security and National Resources News Services, August 15, 2003, http://www.publicedcenter.org/stories/trento/2003-08-15.

48. Priest and Farah, "Terror Alliance Has U.S. Worried."

49. Marc Perelman, "Brazil Connection Links Terrorist Groups," *Forward*, March 21, 2003.

50. Henry Orrego, "Bin Laden Trail Grows Cold on South America's Triple Frontier," Al-Jazeera, May 20, 2003.

51. Ibid.

52. Ibid.

53. Sebastian Junger, "Terrorism's New Geography," *Vanity Fair* (December 2002).

54. Marc Perelman, "Brazil Connection Links Terrorist Groups." Also see James Robbins, "Ambassadors of Terror," *National Review*, April 10, 2003.

55. Jim Bronskill, "Terrorists Eyed Ottawa Targets," *Ottawa Citizen*, February 8, 2004.

56. Ibid.

57. Neil Mackay, "After Egypt, Where Will Al Qaeda Strike Next?" *Sunday Herald*, October 10, 2004.

58. George Jahn, "Brazil Has Tentatively Agreed to Let U.N. Atomic Watchdog View Parts of Its Equipment to Enrich Uranium," Associated Press, October 6, 2004.

59. Con Coughlin, *Saddam: King of Terror* (New York: HarperCollins, 2002), p. 137.

60. Reese Ewing, "Brazil Police Seize Black Market Uranium Ore," Reuters, August 25, 2004.

61. Dave Emory, "Three's a Crowd: Terrorism and the Triple Border Area," *For the Record*, March 25, 2004.

62. Craig Unger, "Saving the Saudis," *Vanity Fair* (October 2003).

63. David Kalish, "Charming and Well-Connected, Bin Laden Family Spans the Globe," Associated Press, October 4, 2001.

64. Unger, "Saving the Saudis."

65. Andersen, "Al Qaeda Across America."

66. Stakelbeck, "Terror's South American Front."

67. Ibid.

68. Deroy Murdock, "A Latin American Axis," *National Review*, October 4, 2002.

69. "US 'Sure' of Brazil's Nuclear Plans," BBC News, October 5, 2004.

70. Gary Milhollin and Liz Palmer, "Brazil's Nuclear Program," *Science* (October 2004).

71. Jahn, "Brazil Has Tentatively Agreed to Let U.N. Atomic Watchdog View Parts of Its Equipment to Enrich Uranium."

72. Ibid.

73. Louis Charbonneau, "Did Brazil Buy a Black Market Nuke?" Reuters, September 30, 2004.

74. Constantine Menges, quoted in Murdock, "A Latin American Axis."

CHAPTER NINE: TOO LITTLE, TOO LATE

1. Rohan Gunaratna, *Inside Al Qaeda: Global Network of Terror* (New York: Berkley Books, 2002), p. 221.

2. Richard Labevieve, *Dollars for Terror* (New York: Algora Publishing, 2002), pp. 224–25.

3. John C. K. Daly, "The Latin Connection," *Terrorism Monitor* (Jamestown Foundation) 1, no. 3 (October 10, 2003).

4. Ibid.

5. Rex Hudson, "Terrorist and Organized Crime Groups in the Tri-Border Area of South America," A Report Prepared by the Federal Research Division, Library of Congress, under an Interagency Agreement with the US Government, July 2003.

6. "A Global Overview of Narcotics-Funded Terrorist and Other Extremist Groups," Federal Research Division, Library of Congress under an Interagency Agreement with the Department of Defense, May 2002.

7. Daly, "The Latin Connection."

8. "A Global Overview of Narcotics-Funded Terrorist and Other Extremist Groups."

9. Ibid.

10. Thomas Muirhead, "Terrorism's Triple-Border Sanctuary: Islamists Organize World Terror from Bases in Argentina, Brazil, and Paraguay," *Global Politician*, December 13, 2004.

11. Rex Hudson, "Terrorist and Organized Crime Groups in the Tri-Border Area."

12. Ibid.

13. Jim Bronskill, "Terrorists Eyed Ottawa Targets," *Ottawa Citizen*, February 8, 2004. Also see Martin Arostegui, "Search for Bin Laden Looks South," United Press International, October 12, 2001.

14. Arostegui, "Search for Bin Laden Looks South."

15. Jeffrey Goldberg, "In the Party of God," *New Yorker*, October 28, 2001.

16. Erick Stakelbeck, "Terror's South American Front," *Front Page*, March 19, 2004, http://www,frontpagemag.com/Articles/Read Article.asp ?ID=12643.

17. Sebastian Junger, "Terrorism's New Geography," *Vanity Fair* (December 2002).

18. Hudson, "Terrorist and Organized Crime Groups in the Tri-Border Area of South America."

19. Daly, "The Latin Connection."

20. Peter Hudson, "There Are No Terrorists Here," *Newsweek*, November 19, 2001.

21. Antonio Garrastazu and Jerry Hoar, "International Terrorism: The Western Hemisphere Connection," International Consortium for Alternative Academic Publications, 2003.

22. Joseph Farah, "Al Qaeda South of the Border," *G2 Bulletin*, posted February 18, 2004, http://www.worldnetdaily.com/news/article.asp?Article _ID=37133.

23. Rohan Gunaratna, *Inside Al Qaeda: Global Network of Terror* (New York: Berkley Books, 2002), p. 219.

24. Ibid., p. 220.

25. Marc Perelman, "Feds Call Chile Resort a Terror Hot Spot, *Forward*, January 2, 2003, http://www.forward.com/issues/2003/03.01.03/news2.html.

26. Martin Arostegui, "Search for Bin Laden Looks South," United Press International, October 12, 2001.

27. Ibid.

28. "Narco-News Editorial on Citigroup and White Collar Terrorists, *Narco News*, October 15, 2001, http://www.narconews.com/issue14/white collarterror1.html.

29. "Report Evaluates Al Qaeda Risks World-Wide," *USA Today*, November 11, 2003.

30. "Texaco Faces $1 Billion Lawsuit," BBC News, October 22, 2003.

31. "US Lawyers Say Chevron Texaco Fears Lawsuit," *San Jose Mercury News*, October 24, 2003.

32. Rachel Ehrenfeld, "Terrorism's Drug Money," American Center for Democracy, January 8, 2004.

33. Arostegui, "Search for Bin Laden Looks South."

34. John Otis, "U.S. Anti-Terror Crackdown May Hit Colombia Groups," *Houston Chronicle*, September 25, 2001.

35. Ibid.

36. Bryan Bender, "US Finds a Palatable Word for Military Aid to Colombia," *Boston Globe*, May 6, 2002.

37. "Latin American Security Challenges," Newport Papers, Naval War College, New Port, Rhode Island, 2004.

38. Farah, "Al Qaeda South of the Border."

39. Ivan G. Osorio, "Chavez's Bombshell," *National Review*, January 8, 2003.

40. Ibid.

41. Dale Hurd, "Terrorism's Western Ally," Christian Broadcasting Network, April 22, 2003.

42. Arostegui, "Search for Bin Laden Looks South."

43. "Latin American Security Challenges." Also see "Increased Terrorism in Northern Venezuela," *US News & World Report*, October 6, 2003.

44. Osorio, "Chavez's Bombshell."

45. "Bio Weapons Lab in Venezuela for Saddam and Castro," *Militares Democraticos*, January 7, 2003, http://militaresdemocraticos.surebase.com/articulos/en/20030106-03.html.

46. Ibid.

47. Hurd, "Terrorism's Western Ally."

48. Gordon Thomas, "The Secret War," *Sunday Express*, November 9, 2003.

49. Richard Gott, *In the Shadow of the Liberator: Hugo Chavez and the Transformation of Venezuela* (London: Verso, 2000), p. 58.

50. Johan Freitas, "Chavez Spokesman Reaffirms Support for North Korea," *Militares Democraticos*, January 7, 2003, http://militaresdemocraticos.surebase.com/articulos/en/20030106-03.html.

51. Brigadier General Gonzalez, quoted in ibid.

52. "Latin America on Alert for Terror," *USA Today*, August 21, 2004.

53. "Al Qaeda Said to Recruit in Latin America," *Newsmax*, August 23, 2004, http://www.newsmax.com/archives/2004/8/22/13453.shtml.

54. Honduran Minister Oscar Alvarez, quoted in Julio Medina Murillo, "Analysis: Al-Qaeda Recruiting Hondurans?" *Insight*, January 17, 2005,

http//www.insightmag.com/news/2004/05/11World/Analysis.AlQaeda.Rec
ruiting.Hondurans.html.

55. Ibid.

56. "Latin America on Alert for Terror."

57. "Al Qaeda Operatives Captured by CIA Provided Intelligence
behind New Orange Heightened Terrorist Threat Alert," *Homeland Security
Today*, August 5, 2004.

58. Ibid.

Chapter Ten: The Terrorists and the Gangbangers

1. Matthew Brzezinski, "Hillbangers: MS-13—The Gang with Ties to
El Salvador Moves to the Suburbs," *New York Times*, October 6, 2004.

2. Ginger Thompson, "Latino Gangs Swarm Back to U.S.," *New York
Times*, September 26, 2004.

3. Al Valdez, "Mara Salvatrucha: A South American Import," National
Alliance of Gang Investigators' Association, Orange County District
Attorney's Office, Orange County, California, 2000.

4. Francesco Gomez, "Rising Mara Presence Causes Alarm at PGR,"
Herald (Mexico edition), Mexico City, April 24, 2004.

5. Statement of Thomas A. Constantine, administrator, Drug Enforcement Administration, DEA Congressional Testimony, Senate Committee
on Banking, Housing and Urban Affairs, March 28, 1996.

6. Jerry Seper, "Al Qaeda Seeks Tie to Local Gangs," *Washington Times*,
September 30, 2004. Also see Heather MacDonald, *Crime and the Illegal
Alien* (Washington, DC: Center for Immigration Studies, June 2004).

7. Testimony of Det. Randy A. Merritt, Pasadena Police Department,
before the US House of Representatives, Committee on Government
Reform, Subcommittee on Criminal Justice, Drug Policy, and Human
Resources, November 10, 2003.

8. Valdez, "Mara Salvatrucha: A South American Import."

9. Brzezinski, "Hillbangers."

10. Ibid.

11. Greg Campbell, "Death by Deportation," *Boulder Weekly*, May 27,
2004.

12. Brzezinski, "Hillbangers."

13. Ibid.

14. Molly Shen, "Arrest Made in Deadly Attack on Two Nuns," KGW-TV, Seattle, Washington, September 2, 2002. Also see Michelle Malkin, "INS: Just Following 'Standard Procedure,'" *Jewish World Review*, October 30, 2002.

15. Heather MacDonald, *The Illegal Alien Crime Wave* (New York: Manhattan Institute, January 15, 2004).

16. Valdez, "Mara Salvatrucha: A South American Import."

17. Arian Campos Flores, "Gangland's New Face," *Newsweek*, December 8, 2003.

18. City Attorney Rocky Delgadillo, News Release, Office of the City Attorney, Los Angeles, California, March 25, 2004.

19. Ibid.

20. Mark Stevenson, "Crackdown on Gangs Brings Mexico Violence," Associated Press, December 10, 2003.

21. Donald Bartlett and James Steele, "Who Left the Door Open?" *Time*, September 20, 2004.

22. Bill Hutchinson, "U.S. Borders Porous," *New York Daily News*, September 13, 2004.

23. Bartlett and Steele, "Who Left the Door Open?"

24. http://www.numbersusa.com/hottopic/clearact.html.

25. "ICE Announces Pilot Program to Atlanta and Denver to Reduce Illegal Aliens," Press Release, US Department of Homeland Security, September 28, 2004.

26. Committee Statement of Senator Jeff Sessions, "State and Local Authority to Enforce Immigration Law: Evaluating a Unified Approach to Stopping Terrorists," April 22, 2004.

27. "Al Qaeda Wants to Smuggle N-Material to U.S.," *Nation*, November 17, 2004.

28. Jerry, Seper, "Al Qaeda Leader Identified in 'Dirty Bomb' Plot," *Washington Times*, October 5, 2004.

29. Investigators on Eyewitness News 4, "Terrorist Alley: Illegals from Terrorist Nations Are Crossing the Border into Arizona," KVOA-TV, Tucson, Arizona, aired August 13, 2004.

30. Emma Perez-Trevino, "Potential Terrorists Released Due to Lack of Jail Space, Congressman Says," *Brownsville (TX) Herald*, July 23, 2004.

31. Steve McCraw, quoted in ibid.

32. "Driver's Licenses for Illegals Spark Security Concerns," Fox News, September 8, 2004.

33. Minister Oscar Alvarez, quoted in "Honduran Official: Al Qaida Recruits Central American Gangs," Associated Press, October 21, 2004

34. Investigators on Eyewitness News 4, "Terrorist Alley."

35. Bill Gertz and Rowan Scarborough, "Inside the Ring," *Washington Times*, November 23, 2001.

36. Ibid.

37. "Boston Terror Plot Suspect in Custody," Fox News, January 22, 2003.

38. "Newly Obtained OTM and Special Interest Alien Information," news release, Office of Congressman Tom Tancredo, Colorado's Sixth District, August 10, 2004.

39. Perez-Trevino, "Potential Terrorists Released Due to Lack of Jail Space, Congressman Says."

40. Jerry Seper, "Sensitivities 'Key in Future Arrests,'" *Washington Times*, July 4, 2004.

41. Rep. Solomon P. Ortiz, quoted in Perez-Tevino, "Potential Terrorists Released Due to Lack of Jail Space, Congressman Says."

42. Keach Hagey, "Jamaica Man Pleads Guilty to Giving Al Qaeda Money Supplies," *Queens Chronicle*, August 19, 2004. Also see Elaine Shannon and Tim McGirk, "What Is This Man Planning?" *Time*, August 23, 2004.

43. Julian Coman, *Daily Telegraph* (UK), August 15, 2004.

44. Sheriff D'Wayne Jenigan, quoted in Karen Gleason, "Sheriff's Protest Delays Release," *Del Rio (TX) News Herald*, July 12, 2004.

45. Asa Hutchinson, quoted in Jerry Seper, "Rounding Up Illegals 'Not Realistic," *Washington Times*, September 10, 2004.

46. Sergio Chapa, "Bangladeshi's Arrest Prompts Concern over Border Security," *Brownsville (TX) Herald*, December 10, 2004.

47. Ibid.

48. Michele McPhee, "Arrested MS-13 Member Wanted in Police Slayings," *Boston Herald*, January 12, 2005.

49. Michele McPhee, "East Boston Gang Linked to Al Qaeda," *Boston Herald*, January 5, 2005.

50. Andy Newan and Daryl Khan, "Brooklyn Mosque Becomes Terror Icon," *New York Times*, March 9, 2003.

51. Daniel Eggen and Manuel Roig-Franzia, "FBI on Global Hunt for Al Qaeda Suspect," *Washington Post*, March 21, 2003.

52. Kelli Arena and Kevin Bohn, "Link between Wanted Saudi Man and 'Dirty Bomb' Suspect," CNN, March 22, 2003.

53. Shannon and McGirk, "What Is This Man Planning?"

54. FBI Alert, March 20, 2003.

55. David Kidwell, "Broward Man Sought as Terror Suspect," *Miami Herald*, March 21, 2003.

56. "FBI Manhunt Targets Al Qaeda Suspects," CBS News, March 26, 2004.

57. Bill Gertz, "Al Qaeda Pursued a 'Dirty Bomb,'" *Washington Times,* October 17, 2003. Also see Paul L. Williams, *Osama's Revenge: The Next 9/11—What the Media and the Government Haven't Told You* (Amherst, NY: Prometheus Books, 2004), p. 89; "State Department Offers $5 Million Reward for Al-Qaeda Dirty Bomb Plotter Who Attempted to Enter U.S.," *Nuclear Threat Initiative,* October 5, 2004.

58. Gertz, "Al Qaeda Pursued a 'Dirty Bomb.'" Also see John Loftus, "180 Pounds of Nuclear Material Missing in Canada," WABC, November 7, 2003.

59. "Most Wanted: The Next Atta," *60 Minutes,* CBS News, March 26, 2004.

60. James Gordon Meek, "Officials Fear Al Qaeda Nuclear Attack," *New York Daily News,* March 14, 2003.

61. Ibid. Also see Williams, *Osama's Revenge: The Next 9/11*, pp. 88–89.

62. Meek, "Officials Fear Al Qaeda Nuclear Attack." Also see Elaine Shannon and Michael Weisskopf, "Khalid Sheikh Mohammed Names Names," *Time,* March 24, 2003.

63. Meek, "Officials Fear Al Qaeda Nuclear Attack."

64. "FBI Seeking Public Assistance in Locating Individuals Suspected of Terrorist Activities," FBI National Press Office, March 20, 2004. Also see Evan Thomas, David Klaidman, and Michael Isikoff, "Enemies among Us," *Newsweek,* June 7, 2004.

65. Jim Kirksey, "Two Suspected Al Qaeda Agents Dropped In for Meal, Says Denny's Manager," *Denver Post,* May 28, 2004.

66. Shannon and McGirk, "What Is This Man Planning?"

67. "Al Qaeda Said to Recruit in Latin America," Associated Press, August 22, 2004.

68. Ibid.

69. "Border Breach Stirs Fears," *Dallas Morning News,* August 14, 2004.

70. Michael Marizco, "Sonora on Alert for #1 Al Qaida Suspect," *Arizona Daily Star,* August 18, 2004.

71. Luke Turf, "Al Qaida Leader May Try to Cross Border," *Tucson Citizen,* August 18, 2004.

72. "Al Qaeda Wants to Smuggle N-Material to US," *Nation,* November 17, 2004.

73. Anna Cearley and Ornell R. Soto, "Reason for Plane Theft Worrisome," *San Diego Union Tribune,* November 17, 2004.

CHAPTER ELEVEN: THE SLEEPER CELLS

1. Steven Emerson, "Terrorism Financing and US Financial Institutions," testimony before the House Committee on Oversight and Investigations of Financial Services, March 11, 2003.

2. Jason Williams and Andrew Brent, "The World Comes to Atlantic Avenue," *Street Level*, New York University School of Journalism, July 15, 2003, http://journalism.nyu.edu/publiczone/streetlevel/atlanticave/world/money.htm.

3. Paul L. Williams, *Al Qaeda: Brotherhood of Terror* (Indianapolis: Alpha Books, 2002), p. 10.

4. Daniel Benjamin and Steven Simon, *The Age of Sacred Terror* (New York: Random House, 2002), p. 5.

5. Nosair, quoted in ibid.

6. John Miller, "A Decade of Warnings: Did Rabbi's 1990 Assassination Mark Birth of Islamic Terror in America?" *20/20*, ABC News, August 16, 2002.

7. Carl Limbacher, "Mosque Linked to '93 World Trade Center Bombing Funded Bin Laden," *Newsmax*, November 26, 2003.

8. Peter Lance, *100 Years of Revenge* (New York: Regan Books, 2003), pp. 38–42.

9. Abdullah Azzam, quoted in Peter L. Bergen, *Holy War, Inc.* (New York: Simon & Schuster, 2002), p. 136.

10. "Indictments Were Just Not Enough," *Newsday*, May 7, 2004.

11. Daniel Pipes, *Militant Islam Reaches America* (New York: Norton, 2003), p. 137.

12. Ibid.

13. Bergen, *Holy War, Inc.*, p. 138.

14. Simon Reeve, *The New Jackals: Ramzi Yousef, Osama Bin Laden, and the Future of Terrorism* (Boston: Northeastern University Press, 2002), p. 24.

15. Benjamin and Simon, *The Age of Sacred Terror*, p. 7.

16. John Miller, "A Decade of Warnings."

17. Sheikh Omar Abdel Rahman, quoted in Benjamin and Simon, *The Age of Sacred Terror*, pp. 16–17.

18. Ibid., p. 19.

19. Joe Kaufman and Beila Rabinowitz, "Father Knows Terror Best," *Front Page*, October 27, 2003, http://www.frontpagemag.com/articles/ReadArticle.asp?ID=10517.

20. Ibid.

21. Harvey Kushner and Bart Davis, *Holy War on the Home Front* (New York: Sentinel, 2004), p. 64.

22. Marina Jimenez, "The Radicalization of US Moslems," *National Post*, November 17, 2001.

23. Ibid.

24. Ibid.

25. Pipes, *Militant Islam Reaches America*, p. 123.

26. Dave Eberhart, "Muslim Moderate Kabbani Firm on Terrorist Nuclear Threat," United Press International, *Newsmax.com*, November 19, 2001, http://www.newsmax.com/archives/article.asp?Article_ID=29100.

27. Mark Clayton, "How Are Mosques Fighting Terror?" *Christian Science Monitor*, August 12, 2002.

28. Kushner and Davis, *Holy War on the Home Front*, p. 69.

29. "Muslim American Politics after September 11: Transcript of the Center's Conversation with Ahmed H. al-Rahim," December 29, 2003.

30. The President's State of the Union Address, *CNN.com*, June 29, 2002, http://www.cnn.com/2002/ALLPOLITICS/01/28sotu.transcript.

31. *Wolf Blitzer Reports*, CNN, December 14, 2001.

32. Pipes, *Militant Islam Reaches America*, p. 146.

33. Shaykh Hisham Kabbani, "Islamic Extremism: A Viable Threat to U.S. National Security," report published by the US Department of State, January 7, 1999.

34. Williams, *Al Qaeda: Brotherhood of Terror*, p. 10.

35. Kushner and Davis, *Holy War on the Home Front*, pp. 6–7.

36. Sarah Downey and Michael Hirsch, "A Safe Haven," *Newsweek*, September 30, 2002.

37. Kushner and Davis, *Holy War on the Home Front*, p. 70.

38. Moussaoui, quoted in ibid., p. 71.

39. *Al Qaeda Training Manual*, US Department of Justice, released on December 7, 2001, in accordance with the Freedom of Information Act.

40. Ibid.

41. Ibid.

42. Ibid.

43. Bergen, *Holy War, Inc.*, p. 130.

44. Ibid.

45. Peter Waldman, "The Infiltrator: Ali Mohamed Served in the U.S. Army and Bin Laden's Circle," *Wall Street Journal*, November 26, 2001.

46. Ibid.

47. Tom Tarnipseed, "A Continuum of Terror—from *Mujahadeen* to Al

Qaeda," Common Dreams News Center, November 28, 2001, http://www
.commondreams.org/views01/1128-10htm.

48. Waldman, "The Infiltrator."

49. Paul L. Williams, *Osama's Revenge: The Next 9/11—What the Media
and the Government Haven't Told You* (Amherst, NY: Prometheus Books,
2004), p. 152.

50. Waldman, "The Infiltrator."

51. Pipes, *Militant Islam Reaches America*, p. 147.

52. Nabil Sharef, quoted in Waldman, "The Infiltrator."

53. Steven Emerson, executive director of the investigation, testimony
before the US Senate's Committee on the Judiciary, December 4, 2001.

54. *The United States v. Osama Bin Laden et alia*, United States District
Court, Southern District of New York, May 2, 2001.

55. Steve Emerson, testimony. Also see Daniel Pipes, "Usama Bin
Laden and Herndon, Virginia," *Jewish World Review*, June 19, 2003.

56. John Miller, "Interview with Osama Bin Laden," *Nightline*, ABC
News, June 8, 1998.

57. Pipes, *Militant Islam Reaches America*, p. 149. See also news briefs in
the *Tampa Tribune*, June 10, 2001, and the *Jerusalem Post*, June 20, 2001.

58. This is the way Abdel-Hafiz was described by ABC News. See
Michael Isikoff and Mark Hosenball, "Reinstated," *Newsweek*, February 25,
2004.

59. Marlena Telvick, "The Story of Gamal Abdel-Hafiz: Former Agent in
the FBI's International Terrorism Squad," *Frontline*, Public Broadcasting
System (PBS), September 11, 2002.

60. Daniel Pipes, "The FBI Fumbles," *New York Post*, March 14, 2003.

61. Gamal Abdel-Hafiz, quoted in ibid.

62. Ibid.

63. Ibid.

64. Mark Flessner, quoted in Telvick, "The Story of Gamal Abdel-Hafiz:
Former Agent in the FBI's International Terrorism Squad."

65. John Vincent, quoted in Pipes, "The FBI Fumbles."

66. Ibid.

67. "FBI Charges Florida Professor with Terrorist Activities," CNN, February 20, 2003.

68. Ibid.

69. Michael Fechter, "FBI Agent Who Refused to Tape al-Arian Is Sus-
pended," *Tampa Tribune*, March 3, 2003.

70. Isikoff and Hosenball, "Reinstated."

71. Valerie Kalfrin, "New Indictment Could Delay al-Arian Trial," *Tampa Tribune*, October 6, 2004.
72. Graham Brink, "Al-Arian's Trial Set for Early 2005," *St. Petersburg Times*, June 6, 2003.
73. Isikoff and Hosenball, "Reinstated."
74. Michael Isikoff, "Tensions in the FBI: Why Was This Agent Fired?" *Newsweek*, October 20, 2004.
75. Pipes, "The FBI Fumbles."
76. Isikoff and Hosenball, "Reinstated."
77. Telvick, "The Story of Gamal Abdel-Hafiz."
78. Isikoff and Hosenball, "Reinstated."
79. Ibid.
80. "Current Time," *Bulletin of Atomic Scientists* 42, no. 17 (March 25, 2003).

CHAPTER TWELVE: AMEN, AMERICA

1. Gen. Eugene Habiger, quoted in Graham Allison, *Nuclear Terrorism: The Ultimate Preventable Catastrophe* (New York: Times Books/Henry Holt and Company, 2004), p. 6.
2. James Carroll, "For Nuclear Safety, the Choice Is Clear," *Boston Globe*, October 26, 2004. Also see Mike Allen, "Chemical, Nuclear Arms Still Major Threat, Cheney Says," *Washington Post*, December 17, 2003.
3. "Ashcroft: Nuclear Bombs Biggest Threat in Terror War," *Columbia Daily Tribune*, January 28, 2005. Also see Tom Ridge in Allison, *Nuclear Terrorism*, p. 6.
4. Ashcroft's letter of resignation in "Attorney General Quits the US Cabinet," BBC News, November 10, 2004.
5. "Warren Buffett Warns of Terror Risks," BBC News, May 6, 2002.
6. Graham Allison, "Could the Worst Be Yet to Come?" *Economist*, November 1, 2001.
7. Bill Keller, "Nuclear Nightmares," *New York Times Magazine*, May 26, 2002.
8. "Interview with Michael Scheuer," *60 Minutes*, CBS News, November 14, 2004.
9. Ibid.
10. Allison, *Nuclear Terrorism*, p. 52.
11. Ibid.

12. Wilfred Burchett, quoted in Jack Geiger, "The Mushrooming Cloud," *Nation*, July 13, 1998.

13. Keller, "Nuclear Nightmares."

14. Stuart Taylor Jr., "A Nuclear Nightmare: It Could Happen Here," *Atlantic*, November 14, 2001.

15. Keller, "Nuclear Nightmares."

16. Daniel Benjamin and Steven Simon, *The Age of Sacred Terror* (New York: Random House, 2002), p. 398.

17. Josh Meyer and Greg Krikorian, "US Braces for Attacks by Al Qaeda," *Los Angeles Times*, December 24, 2003.

18. Bill Nichols, Mimi Hall, and Peter Eisler, "'Dirty Bomb Threatens U.S. with Next Terror Attack," *USA Today*, June 11, 2002.

19. Ibid.

20. Ibid.

21. "Dirty Bombs: Response to a Threat," Public Information Report, *Journal of the Federation of American Scientists* (March/April 2002).

22. Ibid.

23. Dr. Henry Kelly, testimony before the Senate Subcommittee on Foreign Relations, March 6, 2002.

24. Official, quoted in Allison, *Nuclear Terrorism*, p. 8.

Epilogue

1. "Doomsday Clock Set Closer to Armageddon," BBC News, February 27, 2002.

2. Bill Keller, "Nuclear Nightmares," *New York Times Magazine*, May 26, 2002.

INDEX

highly enriched uranium (HEU),
103
See also nuclear weapons
US Army and al Qaeda sleepers,
129, 173, 183–85
US Central Intelligence Agency, 28,
94, 110, 141, 142, 192, 194,
214
aid to Taliban, 35–36, 174
US Congressional Task Force on
Terrorism and Unconventional
Warfare, 57, 93
US Department of Counterter-
rorism, 54
US Federal Bureau of Investigation,
142, 172, 184–85, 186–89,
261n58
US Homeland Security, 160, 192
US House of Representatives, 171
Armed Services Subcommittee on
Military Research, 100
Armed Services Subcommittee on
Readiness, 162
Committee on Government
Reform, 100
Select Committee on Homeland
Security, 177–78
Western Hemisphere Subcom-
mittee, 119
US Immigration and Naturaliza-
tion Service, 177
US Intelligence Committee, 120
US Marine Corp, barracks bombed,
126
US National Regulatory Commis-
sion, 198
US Naval War College, 139
US Strategic Weapons, 192
Uthman, Sami, 228

UTN. *See* Ummah Tameer E-Nau
(UTN)

Vahidi, Ahmad, 130
Valindaba nuclear facility (South
Africa), 28
Vasconcelos, Jose Luis Santiago,
149
Vega (newspaper), 132
Venezuela, 146–49
Directorate for Intelligence, Secu-
rity and Prevention (DISIP),
147
Vincent, John, 188
Volynkin, Igor, 98
Vulgar Betrayal, 187
VX nerve gas, 72, 230

Wahhab, Muhammad ibn Abdul,
30
Wahhabist school, 30, 109, 178,
194
Wahhaj, Siraj, 171
Walid, Mafouz Ould, 40
weapons of mass destruction, al
Qaeda's search for, 207–31
Weldon, Curt, 92, 100
Whitcraft, Terri, 92
White, Tony, 64
"Why We Fight America" (Ghaith),
15–21
Woolsey, James, 139
World Trade Center bombings. *See*
bombings by al Qaeda
Wright, Robert, 172

Yablokov, Aleksey, 90
Yayyar, Jafar At. *See* Shukrijumah,
Adnan el-